T0235875

INTERNATIONAL CENTRE FOR MECHANICAL SCIENCES

COURSES AND LECTURES - No. 284

ALGORITHM DESIGN

FOR

COMPUTER SYSTEM DESIGN

EDITED BY

G. AUSIELLO

UNIVERSITA' DI ROMA

M. LUCERTINI

UNIVERSITA' DI ROMA

P. SERAFINI

UNIVERSITA' DI UDINE

SPRINGER-VERLAG WIEN GMBH

This volume contains 52 illustrations

This work is subject to copyright.

All rights are reserved,

whether the whole or part of the material is concerned

specifically those of translation, reprinting, re-use of illustrations,

broadcasting, reproduction by photocopying machine

or similar means, and storage in data banks.

© 1984 by Springer-Verlag Wien

Originally published by Springer Verlag Wien New York in 1984

ISBN 978-3-211-81816-9   ISBN 978-3-7091-4338-4 (eBook)

DOI 10.1007/978-3-7091-4338-4

PREFACE

PREFACE

For a long time the design of data processing systems has been mainly based on experience and practical considerations more than on formal quantitative approaches. This was true both in the design of computer architecture and in the design of software systems such as operating systems and database systems. Two relevant exceptions were the early studies in switching theory, concerned with such problems as minimization and reliability, on the one hand, and on the other hand, the general mathematical approach to computer system modeling and performance evaluation. More recently, the evaluation of distributed computing related to technological advances in microelectronics, has increased the need for quantitative studies for optimizing the design of computer systems.

As the complexity of computer systems grows, the need for formalization and theoretical analysis is becoming more and more important. The development of formal semantics has provided tools for dealing with correctness and other desirable properties of distributed computing, the development of formal models in different areas (such as distributed system layout, data base design, computer network topology, scheduling and routing) has provided tools for dealing with efficiency and performance optimization; advances in theory of algorithms design and technological increases in computing power have led to the feasibility of the exact or well approximated solution of large scale optimization problems; finally advances in the theory of computing and analysis of algorithms and data structures have led to a new approach to the design of algorithms for the efficient solution of hard problems related to distributed processing systems. Such problems as optimal memory management, optimal design of computer networks and multiprocessor systems, optimal layout of VLSI systems, efficient exploitation of parallel computing systems, optimal management of database schemes, concurrency control, have been thoroughly investigated recently with relevant practical results.

Starting from previous experience in the area of analysis and design of algorithms and their application in combinatorial optimization (a School held in September 1979[1] and a Workshop held in September 1982 [2] both organized by the Department of Computer and System Science of the University of Rome and CISM in Udine) it appeared to be of great relevance in computer science to devote a School to the interactions between formal approaches to computer system design and the theory of algorithms.

Therefore in July 1983, a School on "Algorithm design for computer system design" was then held in Udine under the sponsorship of the International Centre for Mechanical Sciences and the Department of Computer and System Science of the University of Rome, and with the financial support of UNESCO and the Italian Research Council, CNR, through its Committee for Mathematics.

The aim of the School was to provide young postgraduates and junior professionals in Computer Science with an uptodate algorithmic approach to the design and optimization of computer systems. Some of the leading scientists in the field were invited to deliver lectures on the state of the art in the following areas: storage allocation and packing problems (E.G. Coffman), design and implementation of VLSI systems (F. Preparata, C.K. Wong), multiprocessor system design (G. Cioffi), network design (F. Maffioli), concurrency control (D.P. Bovet). Other topics related to basic algorithmic and combinatorial problems in computer system modeling and design were presented by the organizers (C. Papadimitriou, G. Ausiello, M. Lucertini).

This volume presents a collection of unpublished papers referring to some of the issues discussed during the School.

The first part of the volume, devoted to combinatorial problems in computer system design, includes an introduction to the complexity of the exact and approximate solution of combinatorial problems (G. Ausiello), a survey on graph optimization and integer programming models of task assignment problems in distributed systems (M. Lucertini), an extended state of the art paper on approximation algorithms for bin packing (E.G. Coffman, M.R. Garey, D.S. Johnson), a paper on topological network design under multiple non simultaneous demands (M. Lucertini, G. Paletta) and a paper on minimal representation of directed hypergraphs and their applications to data base design (G. Ausiello, A.D'Atri, D. Saccà).

The second part presents papers devoted to specific issues in the optimal design of parallel computer systems and includes an introduction on parallel computer models (G. Ausiello, P. Bertolazzi), a state of the art paper on structural organization of MIMD machines (G. Cioffi), a paper on a new proposal for a VLSI sorter (C.K. Wong) and a selected and annotated bibliography on the theory of VLSI layout (F. Preparata).

G. Ausiello, M. Lucertini, P. Serafini

(1) G. Ausiello, M. Lucertini Eds. "Analysis and design of algorithms in combinatorial optimization", CISM Courses and Lectures N. 266, Springer-Verlag, New York, 1981.

(2) G. Ausiello, M. Lucertini Eds. "Analysis and design of algorithms for combinatorial problems", to appear in Annals of Discrete Mathematics, North Holland, 1984.

CONTENTS

CONTENTS

PART I

COMBINATORIAL PROBLEMS IN
COMPUTER SYSTEM DESIGN

COMPLEXITY OF
EXACT AND APPROXIMATE SOLUTION OF
PROBLEMS. AN INTRODUCTION(*)

Giorgio Ausiello
Dipartimento di Informatica e Sistemistica
Università di Roma "La Sapienza"

Abstract. NP-complete optimization problems are frequently encountered in the optimal design of computer systems, operating systems, databases etc. In this paper a discussion of the basic techniques which lead to the characterization of the complexity of optimization problems is presented. The class of optimization problems which are associated to NP-complete decision problems is then presented and various algorithmic techniques for the approximate solution of such problems are introduced. Finally necessary and sufficient conditions for the approximability of optimization problems are given.

(*) This research has been partially supported by MPI Nat. Proj. on "Theory of algorithms".

1. INTRODUCTION: THE PARADYGM OF COMPLEXITY ANALYSIS

In the design and optimization of computer systems, operating systems, database systems, optimization problems which require exponential time to be solved often occur. This happens in a wide variety of cases: multiprocessor scheduling, task assignment in distributed computing, file assignment in distributed databases, VLSI layout problems, computer network design problems, storage allocation, concurrency control problems etc. Many of these problems will be discussed in other contributions in this volume. For these problems no polynomial time algorithm is known until now and, probably, no such algorithm exists. For this reason these problems are considered to be computationally "intractable" and algorithms for determining their approximate solution have to be designed.

The assumption that a polynomially solvable problem is considered "tractable" (even if it may require time n^{100}) and that a problem which cannot be solved in polynomial time is considered "intractable" (even if it does require a slowly growing exponential time such as $2^{n/100}$ or even non exponential time such as $n^{\log n}$) is a natural, though rough, approximation to the characterization of the computational complexity of a problem. In fact, on one side, if we have no a priori information on the size of the instance of the problem that we have to solve we have no other choice than referring to the asymptotic behaviour if our algorithms: on the other side, the difference between a polynomial running time and an exponential running time is so dramatic that only problems of polynomial complexity would benefit of improvements in computer technology. For example, a very strong improvement in computer efficiency, say 1000 times, would in-

crease ten times the size of the largest instance that we
can solve in one hour of computer time if the running time
is n^3, while such size would be increased only by ten if the
running time is 2^n.

Taking into account the single distribution between
tractable and intractable problems, when given a practical
problem P to be solved we usually take the following para-
dygmatic behaviour:

i)   *Determine complexity* of P by establishing

- *upper bound*, that is amount of computer time
sufficient to solve the problem by means
of some algorithm as a function of the input size,

- *lower bound*, that is amount of computer time
needed to solve the problem by whatever
algorithm, due to the intrinsic difficulty
of the problem.

ii) *If P is tractable* try to find the best possible algorithm from the point of view of the

- *worst case behaviour* or of the

- *average case behaviour*

according to the needs of the application.

iii) *If P has not been recognized to be tractable* (no polynomial algorithm has been found) then check whether P
is NP-complete, that is whether it belongs to the class
which is considered to be the threshold between tractability and intractability. To this end, check whether

    - *P is solvable in polynomial time by means of a*
      nondeterministic algorithm (P ∈ NP)
    - there exists a problem P' which is already known
      to be NP-complete and such that P' *may be*
      reduced to P.

iv) *If P has been recognized to be intractable* (e.g. it has
  an exponential lower bound, or it is NP-complete and,
  hence, probably intractable) then
    - *try to find an ε-approximate algorithm,* that
      is an algorithm which provides a solution
      with a relative error smaller than ε with
      respect to the optimal solution
    - *determine the complexity of the approximate pro-*
     blem

v) *If the approximate problem is also intractable* (e.g.
  even to determine an ε-approximate solution is an NP-
  complete problem)
    - *try to find a heuristic algorithm* which efficien-
      tly provides the exact or a good approximate
      solution sufficiently often
    - *determine efficiency and quality of heuristics*
      in the worst case or in the average.

The various steps of this procedure leading to the characterization of the complexity of the exact and of the approximate solution of an optimization problem require a more
precise comprehension of various concepts which are at the
base of computational complexity. The next paragraph will be
devoted to a brief introduction of such concepts. § 3 contains
the illustration of the most significant complexity classes:
P, NP, PSPACE. In § 4 the notion of NP-hard problem will be
introduced and the polynomial degrees of complexity (particularly the NP-complete degree) will be discussed. In § 5

the concept of -approximation will be defined and the basic
approximation techniques will be presented. Finally § 6 is
devoted to the discussion of the approximability and non approximability of NP-complete optimization problems and to
various necessary and/or sufficient conditions for approximability.

2. BASIC CONCEPTS IN COMPUTATIONAL COMPLEXITY

In order to approach the study and analysis of complexity properties of optimization problems various concepts
have to be made more precise because the results which are
obtained may be havily influenced by the choice of several
factors:

i)   *Machine models and complexity measures*
The first element which has to be defined in order to
perform a complexity analysis is the machine model that we
assume for executing our algorithms and the kind of resource
whose computation is assumed as cost of computation. Some of
the most used models, together with their respective measures,
are:

- *Turing machines* with one tape or many tapes, deterministic or nondeterministic (that is capable of executing one or several transitions at the same time: in
  the first case a computation is essentially a chain of
  configurations, in the second case a tree). The measures which are naturally associated to Turing machines
  are time (number of steps) and memory (largest amount
  of work tape required during a halting computation).
- *Register machines* (or RAMs, *random access machines*)
  similar to real computers, provided with a finite
  number of registers capable of containing arbitrarily

large integers, programmable by means of a naïf
machine language. In this case the resources which
are usually taken into consideration are the number
of elementary operations (uniform cost model, UC-RAM)
or, more realistically, the sum of the costs of elementary operations (logarithmic in the size of oper-
ands: logarithmic cost model, LC-RAM).

- *Interpreters of high level naguages*; in this case we
  assume of expressing our algorithms by means of a
  high level language and in order to evaluate the
  complexity we limit ourselves to counting how many
  times the dominant operations are executed as a function of the input size (e.g. how many comparisons to
  sort n integers).
- *Ad hoc models* suitable for expressing algorithms related to particular computational structures: boolean
  circuits, directed acyclic graphs, straightline programs etc.

Actually among some machine models there are relationship
which allow to derive the cost of solution of a problem in
a model when the cost in another model is known. For example
the following measures are mutually polynomially related:

- time for one-tape Turing machines
- time for multi-tape Turing machines
- time for LC-RAM

Also

- space for deterministic Turing machines
- space for nondeterministic Turing machines

are polynomially related.

On the other side it is not known whether a nondeterministic Turing machine may be simulated by a deterministic
on in polynomial time (as we will see this is one
of the major open problems in computer science)

neither is known whether polynomial space bounded Turing machines are indeed more powerful than polynomial time bounded deterministic or nondeterministic Turing machines. Clearly for the sake of establishing tractability or intractability of problems any of the polynomial time equivalent models is adequate.

ii) *Input size*

One of the elements which may influence the evaluation of the complexity of a problem is the way in which we determine the size of the input. In principle we should take into consideration the *overall length* of the input string. In many applications it happens that we may, equivalently, consider some *parameter of the input size* (number of rows in a matrix, number of nodes in a graph etc.). For example when the imput is a vector of integers $a_1, \ldots, a_n$ the overall length is $n \cdot a_{max}$ but when we assume that a_{max} is always smaller than the largest integer which may be contained in a computer word the complexity may simply be expressed as a function of n. In problems with a numerical input the fact that we consider as input size the length of the input and not its value entails a dramatic difference in the evaluation of the complexity because the value of an integer is exponentially larger than its length. As we will see in the following, problems whose complexity is polynomial in the value of the input (and hence exponential in the length) are called *pseudopolynomial*.

iii) *Type of analysis*

The analysis of the behaviour of algorithms is usually performed by considering how much resource the algorithm requires for a given input of size n, as a function of n and by determining the *asymptotic* growth of such function.

Clearly among inputs of size n we may have the possibility of meeting simpler instances of the problem or more com-

plex ones (e.g. in a sorting problem a vector of n integers
may be already ordered or havily unordered). This variety
of possibilities gives rise to various kinds of analysis.

Let π be a program in a given machine model and let
$\tau_\pi(x)$ be the number of steps required by π on input x.

- *Worst case analysis*: in this case the behaviour of the
  algorithm is analysed with respect to the hardest instance
  for any given n:

$$t_\pi(n) = \max\{\tau_\pi(x) \mid |x| = n\}$$

(where $|x|$ denotes the length of instance x).

- *Average case analysis*: when we assume that all instances
  of size n are equally likely we may consider the average
  behaviour of the algorithm and forget about a few particularly hard but rare instances:

$$t_\pi^A(n) = \frac{\sum_{|x|=n} \tau_\pi(x)}{|\{x \mid |x|=n\}|}$$

- *Probabilistic analysis*: in those (frequent) cases in which
  an average case analysis cannot be precisely determined we
  may limit ourselves to defining random instances $\bar{x}_1, \bar{x}_2, \ldots$
  $\ldots, \bar{x}_n, \ldots$ of the problem of size $1, 2, \ldots, n, \ldots$ and determining the expected behaviour of the algorithm on such in-
  stances:

$$t_\pi^P(n) = E[\tau_\pi(\bar{x}_n)]$$

Once the basic factors of the analysis have been determined we may approach the problem of characterizing the complexity of the given problem. As we observed before such task
is usually based on a worst case analysis and is accomplished

by providing two bounds to the complexity:

i) *Upper bound*: amount of resource g' such that at least
    one program π may solve the given problem P asymptotic-
    ally within such resource bound, that is

$$\exists \pi \exists c \exists n_o \forall n > n_o \left[t_\pi(n) \leq c \cdot g'(n) \right]$$

In this case we say that the complexity of P is $O(g')$

ii) *Lower bound*: amount of resource g" such that given any
    algorithm π for P it requires more resource than g"
    asymptotically, that is

$$\forall \pi \exists c \exists n_o \forall n > n_o \left[t_\pi(n) \geq c \cdot g''(n) \right]$$

In this case we say that the complexity of P is $\Omega(g'')$.
Clearly, the closer g' and g" are, the better the complexity
of P is precisely determined. A classical example in which
the lower bound and upper bound are so close that we can
speak in terms of optimal algorithms is sorting. In this
case, both the lower bound and the upper bound (measured in
terms of comparisons) are essentially n log n and this means
that any algorithm with such worst case performance (e.g.
merge sort, heapsort) is asymptotically optimal.

Unfortunately such desirable situations are somewhat rare.
Especially in the case of optimization problems very frequently we have that the strongest lower bound we have is
quadratic while no algorithm which performs better than exponentially is known. In these cases, hence, the most powerful
technique we have is based on the concepts of complexity
classes for expressing upper bounds and of reductions for
expressing lower bounds. These concepts will be briefly
discussed in the next two paragraphs.

3. COMPLEXITY CLASSES. THE CLASS NP

Given a machine model M, a resource T (e.g. time or space for machine model M) a bound t on the resource T, let t_π be the cost of executing program π in the worst case; a *complexity class* is the set of functions

$$C_t^{M,T} = \left\{ f \mid \text{there exists an integer } n_0 \text{ and a program } \pi \text{ for } f \text{ such that } t_\pi(n) \leq t(n) \text{ for all inputs of size } n > n_0 \right\}$$

When given a problem P we determine an algorithm in M for solving P which runs with a cost bounded by $t(n)$ on inputs of size n (for sufficiently large n), we may say that P belongs to the complexity class C_t. For example, on the base of the upper bound mentioned in the preceding paragraph we know that sorting belongs to the class C_{n^2} or, better, to the class $C_{c\,n\log n}$ for a suitable constant c.

Particular relevance among complexity classes have those classes which may be defined as the union of infinitely many classes. The fact that, under suitable conditions, the infinite union of classes may still be a complexity class is one of the fundamental results of computational complexity theory. Here we simply introduce and discuss some of the most important union classes based on time and space for Turing machines.

i) $P = \bigcup\limits_{k>0} C_{n^k}^{TM,TIME}$ is the class of those problems which

may be solved in polynomial time by means of deterministic Turing machines. According to the preceding observations on the relationships between machine models it is clear that a problem is in P if and only if it may be solved in polynomial time also by means of register machines (with logarithmic cost functions) and

by means of any other "reasonable" machine model, including real computers.

ii) $NP = \bigcup_{k>0} C_{n^k}^{NDTM,TIME}$ is the class of those problems

which may be solved in polynomial time by means of non-
deterministics Turing machines. Given a decision problem, that is the problem of deciding whether a given
string x (representation of an instance of the problem)
belongs to a given set A (the set of instances sharing
a given property) we say that a nondeterministic machine
M solves it (M *accepts* A) in polynomial time if there
exists a polynomial p such that for every $x \in A$ M ac-
cepts x and stops in time $p(|x|)$. Clearly the class P
is contained in the class NP and besides a large class
of combinatorial problems which are not known to be in
P have a particular structure which allows to solve them
in nondeterministic polynomial time. General problems
in operations research and most of the combinatorial
problems that we mentioned at the beginning have this
property: graph partitioning problems, layout problems,
bin packing,scheduling problems etc. For all this problems the search space of solutions is a tree of polynomial depth and the solution may clearly be found in
polynomial time by a nondeterministic branching procedure. A typical example of problem which is in NP and
is not known to be in P is the problem of deciding the
satisfiability of boolean expressions in conjunctive
normal form. In fact, given the expression $(\bar{p} \vee q \vee r)$
$\wedge (\bar{q} \vee \bar{r}) \wedge (p \vee r)$ the search space may be generated as
shown in Fig. 1.
A nondeterministic algorithm generates all possible
truth assignments in only three steps and subsequently,
for every truth assignment it checks whether it sati-

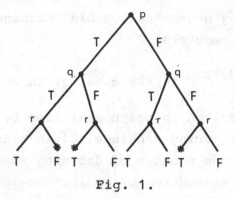

Fig. 1.

sfies the given expression. If we know a method for
simulating a nondeterministic machine by means of a deterministic one in polynomial time, then the classes
P and NP would coincide. Actually the most widespread
conjecture is that such a method cannot exist and that
the classes P and NP are indeed different.

iii) $PSPACE = \bigcup_{k>0} C_{n^k}^{TM,SPACE}$ is the class of those problems

which may be solved within polynomial space by means of
deterministic Turing machines. Since a nondeterministic
Turing machine can be simulated by a deterministic one
within polynomial space we have that the class PSPACE is
also equal to $\bigcup_{k>0} C_{n^k}^{NDTM,SPACE}$. Clearly PSPACE $\supseteq$ NP but

whether such containment is strict or not is again still
an open problem. Typical examples of problems which are
in PSPACE but are not known to be in NP (nor, clearly,
in P) are the problem of deciding whether a given qua-
ntified boolean expression is true, the problem of deciding whether two given regular expressions are equivalent or not, the problem of deciding whether there
exists a forced win for the first player in various
games (such as hex, go, checkers, chess etc.) generalized
on a n × n board.

4. <u>COMPLEXITY DEGREES. THE NP-COMPLETE DEGREE</u>

Unfortunately, given a problem P it is not always possible to determine a neat characterization to its complexity by means of upper and lower bounds. In many problems of practical relevance, such as the problems that we mentioned in the introduction, the best known upper bound and the best known lower bound are $O(2^n)$ and $\Omega(n^2)$ respectively. In such cases the characterization of the complexity may only be achieved in relative terms rather then in absolute terms. By transforming an instance of one problem A into an in instance of another problem B, in fact we may show that the solution of A is at least as hard as the solution of B and in some cases we may show that two problems are equivalently hard.

Let us first consider how these concepts may be formally stated for decision problems. Subsequently we will extend them to optimization problems.

Let two sets A and B be given. We say that

A *is reducible to* B $(A \le B)$ if there exists a many-one function f such that $x \in A$ if and only if $f(x) \in B$. The fact that the problem of deciding whether a string x belongs to a set A may be transformed to the problem of deciding whether the string f(x) belongs to a set B means that, intuitively speaking,

$$\text{complexity}(A) \le \text{complexity}(f) + \text{complexity}(B)$$

If the transformation f is sufficiently simple (that is complexity(f) < complexity(B)) we may say that

$$\text{complexity}(A) \le \text{complexity}(B)$$

More formally, suppose that f is polynomially computable by means of a deterministic Turing machine, then we may say that A is polynomially reducible to B $(A \le_p B)$. Let us see the following example.

Let EXACT-COVER be the set $\{F \mid F$ is a family of sets $S_1, \ldots, S_n \subseteq S = \{e_o, \ldots, e_m\}$ and there exists a subfamily of pairwise disjoint sets $S_{i_1}, \ldots, S_{i_j}$ such that $\overset{j}{\underset{h=1}{\cup}} S_{i_h} = S\}$.

Let SUBSET-SUM be the set $\{\langle a_1, \ldots, a_n, b \rangle \mid$ there exists a 0-1 vector $\bar{x}$ such that $\Sigma a_i x_i = b\}$.

We may show that EXACT-COVER $\leq_p$ SUBSET-SUM by means of the following polynomial reduction: for every i, $1 \leq i \leq n$

$$a_i = \sum_{k \in \{j \mid e_j \in S_i\}} d^k \text{ where } d = n + 1, \text{ and}$$

$$b = \sum_{k=0}^{m} d^k$$

Clearly to every solution of the EXACT-COVER problem there corresponds a solution of the SUBSET-SUM problem and vice-versa, that is if the instance of the SUBSET-SUM problem which is generated by the reduction does not allow a solution, then no exact cover can be found in the family F. For example:

$$F = \{S_1, S_2, S_3, S_4\} \qquad\qquad d = S$$
$$S_1 = \{e_o, e_1\} \qquad\qquad a_1 = 1 + 2$$
$$S_2 = \{e_1, e_2, e_3, e_4\} \qquad\qquad a_2 = d + d^2 + d^3 + d^4$$
$$S_3 = \{e_2, e_3\} \qquad\qquad a_3 = d^2 + d^3$$
$$S_4 = \{e_4\} \qquad\qquad a_4 = d^4$$
$$S = \{e_o, e_1, e_2, e_3, e_4\} \qquad b = 1 + d + d^2 + d^3 + d^4$$

To the solution $F' = \{S_1, S_3, S_4\}$ there corresponds the solution of the SUBSET-SUM problem $a_1 + a_3 + a_4 = b$. As a result of the existence of a polynomial transformation from EXACT-COVER to SUBSET-SUM we may say that the SUBSET-SUM problem is at least

as hard as the EXACT-COVER problem. Besides,since both problems are in NP but for some of them a polynomial algorithm is
known, we may say that if,eventually the EXACT-COVER problem
is shown not to be in NP then also the SUBSET-SUM problem
would be proven not to be in NP while if the SUBSET-SUM problem would be shown to be polynomially solvable so would the
EXACT-COVER problem.

By means of a polynomial reduction we may hence establish
a relative complexity lower bound between two problems.
Actually we may use the concept of reduction to determine an
even stronger relative lower bound: by showing that any problem of a class C may be reduced to a given problem B we may
show that B is at least as hard as the hardest problem in C.
In particular if we consider the complexity class NP we may
give the following definition. A set B is said to be NP-*hard*
if given any set A in NP we have A $\leq_p$ B.

An example of NP-hard problem is the problem
SATISFIABILITY = {w | w is a propositional formula in conjunc-
tive normal form and there exists a truth assignment to pro-
positional variables which satisfies w}.

Such problem has been shown to be NP-hard by means of
the following argument. Let any set A $\in$ NP be given. Let M_A
be the nondeterministic Turing machine which accepts A in
time $p_A(|x|)$ for a suitable polynomial p_A. It is possible to
construct a Boolean expression $w[M_A,x,p_A]$, depending on M_A,
p_A and the input x, whose length is still polynomial in $|x|$
and which represents an accepting computation of M_A on x.
Clearly w can be satisfied if and only if such computation
exists, that is if x $\in$ A.

This result, one of the fundamental results of complexity
theory, shows that SATISFIABILITY is at least as hard as any
other problem in NP. Actually, since we already saw that
SATISFIABILITY is in NP we may say that it is among the hardest problems in NP. When a problem A is in NP and, at the

same time, it is shown to be NP-hard we say that it is NP-
complete. Hence SATISFIABILITY is an NP-complete problem.
When we say that a problem is NP-complete we actually provide
a characterization of its complexity in relative terms; the
membership in NP corresponds in fact to an upper bound while
the NP-hardness corresponds to a lower bound.

The polynomial reducibility among decision problems is
a transitive relation. Two consequences of this property are
particularly relevant. First of all in order to show that a
problem A is NP-hard we may simply show that any other NP-
hard problem (SATISFIABILITY, for example) may be reduced to
A. Secondly, if A and B are both NP-complete problems, since
in this case we have $A \leq_p B$ and $B \leq_p A$ we may say that $A \equiv_p B$,
that is A and B are equivalent in terms of complexity (modulo
a polynomial). The equivalence classes of polynomial reducibility are called *polynomial complexity degrees*. Beside the
NP-complete degree, other examples of polynomial degrees are
the degree of GRAPH-ISOMORPHISM (that is the class of all
those problems whose complexity is equivalent to the complexity of deciding whether two graphs are isomorphic or not)
and, trivially, the class P.

NP-complete problems represent a very interesting class
of probelms. Most problems considered in the introduction
turn out to be in this class. This means that either all of
them may be solved in polynomial time or, more likely, none
of them is. Unless P = NP the only hope we have is to solve
these problems by means of suitable approximate algorithms.
To this issue is devoted the next paragraph.

5. APPROXIMATION ALGORITHMS FOR NP-COMPLETE OPTIMIZATION PROBLEMS

Let us now go back to considering optimization problems.

As we will see the concepts of NP-hardness and NP-completeness may be extended from decision problems to optimization problems.

The fact that many interesting optimization problems are NP-hard and, hence, (probably) computationally intractable has determined the need for various techniques by means of which at least an approximate solution of the given problem may be achieved. When we consider approximation techniques we realize that NP-complete optimization problems fall into different subclasses according to the fact that they may be solved by approximation methods or not.

First of all let us introduce a formalization of the concept of optimization problem. An NP-*optimization problem* is characterized by a polynomially decidable set INPUT of *instances*, a polynomially decidable set OUTPUT of possible *outcomes*, a mapping SOL:INPUT $\rightarrow$ P(OUTPUT) which, given any instance x of the peoblem, nondeterministically provides the *feasible solutions* of x in polynomial time, and a mapping m:OUTPUT $\rightarrow$ N which again in polynomial time provides the *measure* of a feasible solution. We will denote by $m^*(x)$ the best (*maximal* or *minimal*) solution for input x.

To every NP-optimization problem A a decision problem may be associated by considering the set

$$A^C = \{\langle x,k \rangle \mid x \in INPUT \text{ and } k \leq m^*(x)\}$$

Clearly if A is an NP-optimization problem then A^C is a decision problem in NP. If A^C is NP-complete then we say that A is an NP-complete optimization problem (NPCO).

For example, the problem MAX-CLIQUE is an NPCO. It is characterized by the following items:

INPUT = set of (representations of) all finite graphs

OUTPUT= set of (representations of) all finite complete

graphs

SOL(x) = set of (representations of) all complete subgraphs
         of x

m(y)   = number of nodes of y

The associated decision problem is the problem of reco-
gnizing the set CLIQUE = {⟨x,k⟩ | x is (the representation of)
                            a graph which contains a complete
                            subgraph of k nodes}.

Given an NPCO problem we may say that an algorithm A is
an ε-*approximate algorithm* for A if, given any instance
x ∈ INPUT, we have

$$\left| \frac{m^*(x) - m(A(x))}{m^*(x)} \right| \leq \varepsilon$$

that is the algorithm provides a solution with a relative
error smaller than ε. Such an approximate algorithm is said
to provide a *performance guarantee*. This situation is different from the case in which an algorithm in some cases provides
a solution which is optimal, a very close to the optimal,
while in other cases the solution may be arbitrarily far from
the optimal one.

A problem A is said to be *polynomially approximable* if
given any ε > 0 there exists an ε-approximate algorithm for
A which runs in polynomial time. A is said to be *fully polynomially approximable* if A is approximable and there exists
a polynomial q such that given any ε the running time of the
ε-approximate algorithm is bounded by q(|x|,1/ε).

Clearly the fact that a problem A is polynomially approximable is not enough for approaching its solution because
for example it may happen that when we go from the approximation 1/k to the approximation 1/(k+1) the running time of the
approximate algorithm increases from $O(|x|^k)$ to $O(|x|^{k+1})$ and

soon the approximate solution becomes unfeasibly expensive.
When a problem is fully approximable, instead, we may consider it to be essentially (even if not properly) an easy
problem because the running time of the approximate algorithm
does not encrease too much with the required precision.

Let us now examine how approximate algorithms may be
constructed.

A constructive method that for any given ε provides the
corresponding polynomial ε-approximate algorithm A_ε is said
to be a *polynomial approximation scheme* (PAS). If for every
ε the running time is bounded by $q(|x|,1/\varepsilon)$ for some polynomial q we say that the scheme is a *fully polynomial approximation scheme*.

In order to discuss various approximation schemes let
us consider the typical KNAPSACK problem[(*)]. Such problem
may be characterized in the following way:

INPUT = n items $\langle a_1,c_1 \rangle,\ldots,\langle a_n,c_n \rangle$ and bound b

OUTPUT = 0-1 vectors $\langle y_1,\ldots,y_n \rangle$

SOL(x) = 0-1 vectors $\langle y_1,\ldots,y_n \rangle$ such that

$$\sum y_i a_i \leq b$$

m(x) $= \sum y_i c_i$

It consists in choosing a set of items such that the
profit $\sum y_i c_i$ is maximized while the constraint b on the oc-
cupancy is satisfied.

The fundamental technique for constructing fully polynomial approximation schemes are all based on the classic

---

(*) The variation of knapsack problem in which $a_i = c_i$ for
all i is called SUBSET-SUM problem.

dynamic programming scheme. This scheme, in the case of the knapsack problem, can be summarized as follows:

```
L:=∅;
for all items i in x do
        for all sets S_j in L do
            if S_j ∪ {i} satisfies the constraint b
            then
                begin insert S_j ∪ {i} in L;
                      eliminate dominated elements
                end
        end
    end.
```

take the best solution in L.

It is easy to see that the number of steps of the algorithm is proportional to the number of items in x times the lenght of the list L.

Clearly variations of this scheme are obtained by considering different conditions of dominance between elements.

In the case of knapsack we can define the following dominance rule:

Given two sets S_1 and S_2 in L we say that S_1 is *dominated* by S_2

$$\text{if} \quad \sum_{i \in S_1} c_i \leq \sum_{i \in S_2} c_i \quad \text{and}$$

$$\sum_{i \in S_1} a_i \geq \sum_{i \in S_2} a_i.$$

Clearly the elimination of S_1 does not introduce any error.

Therefore we can obtain the following exact algorithm for the knapsack problem:

Algorithm A_1

L:=∅;

for i = 1 *to* n *do*

    for all sets S_j in L *do*

        if $\sum\limits_{j \in S_j} a_j + a_i \leq b$

           then

                begin L:=L ∪ (S_j ∪ {i})

                      eliminate all S' ∈ L

                      such that ∃S" ∈ L

$$\sum\limits_{j \in S'} c_j \leq \sum\limits_{j \in S''} c_j$$

                      and

$$\sum\limits_{j \in S'} a_j \geq \sum\limits_{j \in S''} a_j$$

             end

      end

end

take the best solution in L.

    To evaluate the complexity of the above algorithm it is sufficient to see that, at each step, the number of solutions contained in the list L is less than $\min(b, \sum\limits_{j=1}^{n} a_j, \sum\limits_{j=1}^{n} c_j)$. So with a suitable implementation of the elimination step it is not hard to see that the complexity of algorithm A_1 is $0(n \cdot \min(b, \sum\limits_{j=1}^{n} a_j, \sum\limits_{j=1}^{n} c_j))$, which means a complexity exponential in the size of the input, as we use a binary encoding for the numbers of the input.

    In order to achieve a fully polynomial approximation scheme the first technique which was used for finding an approximate solution to the knapsack problem was based on scaling all coefficients a_i by a factor $K = \varepsilon \cdot a_{MAX}/n$.

This technique is shown by the following algorithm

Algorithm A_2

    for i = 1 *to* n *do*

      $c_j' = k \cdot c_i$

    end;

    Apply algorithm A_1 taking as input

    $(c_1' \ldots, c_n'; a_1 \ldots, a_n; b)$

    take the best solution and multiply

    it for k.

If $m(A_2(x))$ is the value of the approximate solution we have that

$$m^*(x) - m(A_2(x)) \leq n \cdot k$$

On the other side we can assume that

$$m^*(x) \geq c_{MAX}.$$

It follows that

$$\frac{m^*(x) - m(A_2(x))}{m^*(x)} \leq \frac{c \cdot k}{c_{MAX}} = \varepsilon$$

With respect to the running time we have that the complexity of the algorithm is $O(n \cdot (\sum c_i'))$. Due to the scaling we have that

$$\sum c_i' \leq \frac{n \cdot c_{MAX}}{k} = \frac{n^2}{\varepsilon}$$

So the overall complexity is $O(\frac{n^3}{\varepsilon})$.

Such approximation scheme although very useful for many problems, suffers some drawbacks.

In fact in order to find the fully polynomial approximation scheme we need to know good bounds to m^* and this is a severe limitation to the generality of the method as it can be easily seen if we simply switch from max knapsack to min knapsack problems.

Another limitation of this scheme is that it cannot be applied for solving other NP-complete optimization problems which instead can be shown to be fully approximable by other methods such as the product knapsack problem.

Due to these facts the search for general full approximation schemes has been pursued with the aim of finding results which, despite of a slight loss in efficiency, could be applied to a broader class of problems and that could provide some insight in the properties of fully approximable problems and in their characterization.

The first attempt to provide such a general scheme was the *condensation algorithm*. With respect to the dynamic programming scheme (A_1) the elimination step is performed by eliminating more partial solutions and therefore introducing an error.

More precisely we say that S_2 dominates S_1

$$\text{if } (1-\delta) \sum_{i \in S_1} c_i \leq \sum_{i \in S_2} c_i \text{ and}$$

$$\sum_{i \in S_1} a_i \geq \sum_{i \in S_2} a_i$$

where $\delta = \min\{\varepsilon^2, \frac{1}{n^2}\}$, the condensing parameter, is the relative error introduced in the dominance test. As there is a propagation of the error then the total relative error is at least $\delta^2 \leq \varepsilon$. Moreover the running time is $0(\max\{|x^4|, |x^2|/\varepsilon^2\})$.

A different approach which leads to a more efficient algorithm is based on the technique of *variable partitioning* (as opposed to the constant partitioning technique). This method is based on the partitioning of the range of the measure into intervals of exponentially increasing size and on an elimination rule which preserves only one solution for every interval.

To allow a better understanding of the advantages of this approach the method and the results will be given for the knapsack and the product knapsack. It can be immediately extended to other fully approximable problems.

More in detail the method is as follows.

Let R be the range of the possible values of the measure. In a general NP-complete max-subset problem, and therefore in our cases R is smaller than $2^{P(|x|)}$ for some polynomial P and as we will see the whole development of the algorithm allows us to refer only to this general bound without requiring any more precise extimate of a bound for m^*. The range R is then partitioned into K intervals $[0,m_1),[m_1,m_2),$ $\ldots[m_{K-1},m_K)$ where $m_i = (1+\epsilon/n)^i$. Let us denote T_i the i-th interval.

The elimination rule for the 0/1 knapsack is the following:
Given two sets S_1 and S_2, S_1 is *dominated* by S_2 if

if $\quad \sum_{i \in S_1} c_i \in T_i, \; \sum_{i \in S_2} c_i \in T_j$, $j \geq i$ and

$$\sum_{i \in S_1} a_i \geq \sum_{i \in S_2} a_i$$

Clearly changing the sums in products we have the elimination rule for the product knapsack.

In every interval there will be at most one feasible solution and hence, at each iteration, we will have, at most R elements in the list.

The error that may result by using this algorithm may

be bounded as follows. At stage i at most the error $\Delta_i = m_i - m_{i-1}$ may arise; in the worst case this error may happen at every stage. Since there are n stages and since $\Delta_i < \Delta_{i+1}$ we have that $|m^*(x) - m(A_\varepsilon(x))| \le n \, \Delta_{i_{MAX}}$ where i_{MAX} is such that $m_{i_{MAX}-1} \le m^*(x) < m_{i_{MAX}}$. From the above inequalities we deduce that the overall error is

$$\left| \frac{m^*(x) - m(A_\varepsilon(x))}{m^*(x)} \right| \le \frac{n \left[(1 + \frac{\varepsilon}{n})^{i_{MAX}} - (1 + \frac{\varepsilon}{n})^{i_{MAX}-1} \right]}{(1 + \frac{\varepsilon}{n})^{i_{MAX}-1}} = \varepsilon$$

As far as the complexity is concerned, the number of steps of the given algorithm is as usual a function of n and the lenght of the list L. In this case the number of solutions which may be preserved in L is equal to the number of intervals K which should satisfy the following inequalities

$$(1 + \frac{\varepsilon}{n})^K \le 2^{p(|x|)}$$

$$K \log(1 + \frac{\varepsilon}{n}) \le p(|x|)$$

$$K \le \frac{p(|x|)}{\log(1 + \frac{\varepsilon}{n})}$$

Hence with a suitable implementation the complexity of the method is

$$O\left(n \cdot \frac{p(|x|)}{\log(1 + \frac{\varepsilon}{n})}\right)$$

Therefore in the case of knapsack we have that the range R is bounded by $n \cdot a_{MAX}$ and therefore in this case we have a complexity

$$O(n \cdot \frac{\log n + \log a_{MAX}}{\log(1 + \varepsilon/n)}$$

while in the case of product knapsack

$$O(n^2 \cdot \frac{\log a_{MAX}}{\log(1+\varepsilon/n)}.$$

6. APPROXIMABLE AND NON APPROXIMABLE PROBLEMS

As we have already observed, not all NPCO problems are approximable or fully approximable. For many problems it is possible to show that even the problem of determining an approximate solution is intractable.

A classical example of a problem which is not ε-approximable for any ε is the traveling salesman problem (TSP) which consists in determining the shortest cycle which crosses every vertex of a weighted graph exactly once. In order to show that, given any ε, the problem of determining whether there exists an approximate solution with relative error smaller than ε is NP-complete we may use the following reduction from HAMILTONIAN-CIRCUIT.

Let a graph $G = \langle N,A \rangle$ be given. Let us define a complete graph G' with weights on the edges r_{ij}

$$p_{ij} = \begin{cases} 1 + \varepsilon n & \text{if } (i,j) \notin A \\ \\ 1 & \text{if } (i,j) \in A \end{cases}$$

where $n = |N|$.

Clearly an Hamiltonian path in G exists if and only if in G' there exists a traveling salesman four of length n; in fact any other tour would entail a relative error e not smaller than ε

$$e \geq \left| \frac{n+\varepsilon n-n}{n} \right| = \varepsilon$$

Another problems which is known not to be ε-approximable at least for some values of ε is the GRAPH-COLOURING problem; in this case it is known that to find a solution which uses less colours then the double of the chromatic number of the graph is still an NP-complete problem.

Various attempts have been made to characterize the classes of NPCO problems which are not approximable, approximable or fully approximable.

The first characterization is based on the complexity of subproblems of the given problem.

Let an NPCO problem A be given. Let MAX(x) indicate the largest integer which appears in the input to the problem. For example if the input x is a weighted graph MAX(x) indicates the weight of the heaviest edge. Now let us consider the subproblem $\bar{A}_p$ of A obtained by taking into account only those instances in INPUT such that $MAX(x) \leq p(|x|)$ for some polynomial p. We say that A is *pseudopolynomial* if $\bar{A}_p$ is a polynomially solvable problem; A is *strongly NP-complete* of $\bar{A}_p$ is still an NP complete problem. SUBSET-SUM is a clear example of pseudopolynomial problem. In fact since we may solve the SUBSET-SUM problem in time $0(n \cdot b)$ it turns out that the problem is not polynomial in the size of the input, but is polynomial in the value appearing in the input. Hence if we bound MAX by a polynomial function we have a polynomially solvable problem. An the other side the problem MAX-CUT is a strongly NP-complete problem. In fact even if we restrict all weights to be equal to one we still remain with an NP-complete optimization problem. Similarly strongly NP-complete problems are GRAPH-COLOURING, TSP, MAX-CLIQUE etc. Clearly if a problem is strongly NP-complete it cannot be pseudopolynomial and viceversa.

A fundamental result relates pseudopolynomiability and full approximability of problems: if for all input x we have $m^*(x) \leq q(|x|, MAX(x))$ for a given polynomial q, then a problem is fully approximable if and only if it is pseudopolynomial. As a consequence, under the same hypothesis, if a problem is strongly NP-complete it cannot be fully polynomial.

Actually such characterization is not enough. In fact when the condition is not satisfied (as it happens for the PRODUCT-KNAPSACK problem) then the concept of pseudopolynomiability is not necessary to determine the full approximability of a problem.

More recently new conditions have been proposed which completely characterize both approximable problems and fully approximable problems.

In particular on the base of preceding results (appearing in the references) the following more recent result may be shown.

Let us consider the class of optimization problems which may be stated as *subset problems*, that is those problems in which we look for the subsets of a given set of items which satisfy a given property and which maximize (or minimize) a given objective function m. Most problems that we have discussed insofar are indeed subset problems. Given a subset problem P we say that P satisfies an h-*dominance test* if, given any two feasible solutions S_1 and S_2, the fact that

$$\left| \frac{m(S_1) - m(S_2)}{\min(m(S_1), m(S_2))} \right| < h$$

implies that if S_1^* and S_2^* are the best solutions which may be achieved from S_1 and S_2 respectively

$$m(S_1^*) > h \cdot m(S_2^*)$$

A necessary and sufficient condition for the full approximability of a subset problem P is that for a suitable constant
h P satisfies a polynomial h-dominance test.

7. REMARKS AND REFERENCES

The basic concepts of complexity analysis, such as machine
models, types of analysis, upper bounds and lower bounds
are extensively discussed in

- Aho,A.V., J.E.Hopcroft, J.D.Ullman: *The design and analysis
of computer algorithms*, Addison Wesley, 1974.

In the same volume an introduction to the most fundamental
algorithms for searching and sorting, graph problems and
algebraic problems is also given. The definition and the
first examples and properties of NP-complete problems are
presented in two fundamental papers:

- Cook,S.A.: The complexity of theorem proving procedures,
Proc. 3rd Ann. ACM Symp. on Theory of Coputing, 1971.

- Karp,R.M.: Reducibility among combinatorial problems, in
R.E.Muller and J.W.Thatcher (eds.), *Complexity of Computer
computations*, Plenum Press, 1972.

A complete and detailed discussion of various issues concerned with NP-completeness, such as the complexity classes
NP, co-NP, PSPACE, and a long list of the most relevant
NP-complete problems known in various field of mathematics
and computer science are contained in

- M.R.Garey and D.S.Johnson: *Computers and intractability.
A guide to NP-completeness*, Freeman, 1979.

The basic concepts of approximate solution of optimization
problems are also presented in the same volume and in

- Horowitz,E., S.Sahni: *Fundamentals of computer algorithms*, Computer Science Press, 1978.

  A discussion of a large number of optimization problems which may be encountered in the optimal design of computer systems and a presentation of the basic algorithms for their exact or approximate solutions are given in

- Papadimitrou,C.H. and K.Steiglitz: *Combinatorial optimization: Algorithms and complexity*, Prentice Hall, 1982.

  The following papers provide a more precise approach to the characterization of classes of NP-complete optimization problems:

- Paz,A. and S.Moran: NP-optimization problems and their approximation, *Proc 4th Int. Symposium on Automatic, Languages and Programming*, LNCS, Springer Verlag, 1977.

- Ausiello,G. A.Marchetti Spaccamela, M.Protasi: Toward a unified approach for the classification of NP-complete optimization problems, *Theoretical Computer Science*, 12, 1980.

- Ausiello,G., A.Marchetti Spaccamela and M.Protasi: Full approximability of a class of problems over power sets, *6th Colloquium on Trees in Algebra and Programming*, LNCS, Springer Verlag, 1981.

- Korte,B. and R.Schrader: On the existence of fast approximation schemes, Report No. 80163 Institut für Ökonom. und Op. Res., REWU, 1980.

MODELS OF THE TASK ASSIGNMENT PROBLEM IN DISTRIBUTED SYSTEMS

Mario Lucertini
Dipartimento di Informatica e Sistemistica
dell'Università di Roma e Istituto di Analisi dei
Sistemi ed Informatica del
C.N.R., Viale Manzoni 30, 00185, Roma

ABSTRACT

The paper presents a model for optimum partitioning of tasks over a multiple-processor system. The minimization of the interprocessor communications overhead and/or the message average delay are considered as a design criterion. The algorithmic approaches to the problem are briefly described and improvements to the case of multiple copies of tasks are considered. A large set of references covering the area are included.

1. INTRODUCTION

Many papers in the last years on issues on distributed systems have shown the necessity of models, both in the design of the system and in the resources management to avoid underutilization, overhead and congestion.

In the references, a wide range of models concerning optimal partition of objects in distributed systems is listed, see in particular [8, 9,13,14,15,16,20,25,29,46,50,61,65,68]. The area of computer modelling is especially developed for computer networks. Synthesis models: optimization of cost, capacity and lateness of the communication network; synthesis of fault tolerante networks; concurrency control; optimal management policies; analysis (prediction) models: performance evaluation, average transaction response time during peak traffic periods, utilization of various resources and system availability, deadlock detection and avoidance.

One critical design problem of computer systems is that of assigning computational objects (files, programs of different kind) to possibly different nodes in a computer network for query/update/execution purposes.

Many measures of the optimality of the distribution can be considered either as components of the objective function or as constraints of an optimization problem.

One measure of optimality is minimal cost. The cost consist mainly of storage costs, query/update/execution local costs, communication costs and network cost. Unfortunatly, although the model can be as accurate and comprehensive as desired, solution techniques are very complex and wotk effectively only for toy examples.

Another measure of optimality is performance. Common performance objectives are minimum response time and maximum system throughput, but many others objectives can be considered as fault tolerance via alternative routing capabilities and minimum communication flow on the interconnection links or busses. In this framework distributed systems are commonly represented as queueing networks. Its performance is optimized with respect to some parameters (or decision variables) such as: network topology, routing and scheduling strategies, device speeds, computational graph embedding strategies, device visit ratios. In this paper only the placement of computational objects among interconnected processors is considered.

The paper presents a basic model for exact or approximate optimum partitioning and allocation of tasks over multiple-processor nodes. Minimization of the interprocessor communications overhead and/or the corresponding message average delays has been chosen as a design criterium. Improuvements of the basic model for some classes of applications are considered.

For sake of simplicity a completely homogeneous computer system is considered with all processors of equal capabilities and all processors interconnected by a fully connected network. However the model can be easily generalized to networks with given message paths among all node pairs. The important case of random routing capability is not considered in this paper. Furthermore all transportation cost may be considered equal within the network. The performance in term of delay (see objective function OF 2.1) or in term of speed (number of bytes per unit time, see OF 2.2) for interprocessor communications is considered constant. In the first case the communication channels are supposed to have a speed growing with the traffic incident on the channel such that the message delay remain constant. In case of networks requiring multiple interprocessor transmissions for each message, the total delay will be obtained as the sum of the delays on the utilized channels, plus the delays on the intermediate nodes of the path.

In section 2 the basic model is presented; in section 3 some algorithms for the different objective functions presented in section 2 are briefly described; in section 4 some modifications of the basic model for given classes of applications are investigated.

2. THE BASIC MODEL

2.1. Let be given:

- a fully connected network of C similar computer with speed S (number
  of instruction per unit time);

- n computational objects (files, tasks, jobs,...) to be processed by
  the computer network, N_i represents the (average) amount of instructions to be executed to process the computational object i (N_i/S is
  the time needed to process i);

- a computational graph G(N,A) ($|N| = n$, $|A| = m$) where N represents the
  set of computational objects and A the set of communication requirements among the nodes, the arc weights A_{ij} represent the (average)
  number of node j execution requests sent from node i, the node weights
  A_{ii} represent the (average) number of node i execution requests sent
  from outside the network, M_{ij} represents the number of bytes exhanged
  between i and j for each execution request from i to j.

     The delay time analysis is performed under the standard hypotheses
of Poisson arrivals of all execution request, exponential distribution
of service times and independence assumptions [23].
     Introducing the binary decision variables x_{ik} and the (average)
total execution time for each node E_i, defined as:

$$x_{ik} = \begin{cases} 1 \text{ if node i is processed by computer k} \\ \\ 0 \text{ else} \end{cases}$$

$$E_i = (\sum_{j=1}^{n} A_{ji})N_i/S = A_i N_i/S$$

we can easily write the following constraints of the basic optimization
problem.

2.2. *Basic model constraints*

$$\sum_{k=1}^{C} x_{ik} = 1 \qquad \forall i \qquad\qquad (1)$$

$$E(x^k) = \sum_{i=1}^{n} E_i x_{ik} \leq 1 \qquad \forall k \qquad\qquad (2)$$

$$x_{ik} = 0, 1 \qquad \forall (i,k) \qquad\qquad (3)$$

The first set of constraints indicates that each node must be as-signed to a computer, the equality imply that no multiple copies of nodes are allowed, in some applications this assumption is too restrictive and (1) will be relaxed to inequality constraints indicating that each node must be assigned to at least one computer (see section 4).

The second set of constraints indicates that the total computational load assigned to each computer (with the given speed S) must be less or equal to the computer capacity. In fact this set of constraints is mean-ingfull only if we optimize with respect to the interprocessor overhead with deterministic arrivals. If we take into account delay and we have Poisson arrivals the objective function is build such that if the total load allocated to a computer tend to 1 the delays tend to $+\infty$ (a maximum load factor of about .8 is in practice acceptable). The two possibilities are investigated in the following sections.

2.3. *Basic model objective function*

OF1) *Minimization of interprocessor communications overhead*

$$\min\left(\sum_{\substack{k=1 \\ h \neq k}}^{C} \sum_{\substack{h=1 \\ }}^{C} \sum_{i=1}^{n} \sum_{\substack{j=1 \\ j \neq i}}^{n} x_{ik} x_{jh} A_{ij} M_{ij}\right)$$

or equivalently:

$$\text{cost} + \max\left(\sum_{k=1}^{C} \sum_{i=1}^{n} \sum_{\substack{j=1 \\ j \neq i}}^{n} x_{ik} x_{jk} A_{ij} M_{ij}\right)$$

OF2) *Minimization of system delays*

There exists two kind of delays in the system; the first one is the execution delay in the computers both for execution requests from outside the network and for execution requests from other nodes on the same computer or on other computers, the second one is the transmission delay for execution requests coming from nodes in other computers. The first one depends on the computer speed and the computer load, the second on the **transmission** channel speed and the channel load. A realistic hypothesis is that the channel speed depends on the load in such a way to mantain constant the transmission delay; in other words the links among computers are built after the allocation of nodes on the computers in order to meet such requirement. Under this assumption the transmission delay is equal to a constant for each couple of nodes allocated on different computers and the transmission time depends only on the lenght of the data stream to be sent (M_{ij}). The first expression (OF2.1) of the objective function holds under this assumption. Otherwise, if the speed of the intercomputer channels is a given value T (number of bytes per unit time), the objective function is shown in the second expression (OF2.2).

The computer k average execution delay W_k can be obtained utilizing the standard queueing systems formulae:

$$W_k = \frac{1}{\mu_k - f_k}$$

$$\mu_k = S \sum_{i=1}^{n} x_{ik} / \sum_{i=1}^{n} N_i x_{ik}$$

$$f_k = \sum_{j=1}^{n} \sum_{i=1}^{n} A_{ij} x_{jk} = \sum_{j=1}^{n} A_j x_{jk}$$

It is easy to verify that $\mu_k > f_k$ if and only if $\sum_{i=1}^{n} E_i x_{ik} < 1$. The total computer delay D_C is given by:

$$D_C = \sum_{k=1}^{C} f_k W_k$$

The total transmission delay D_T is given by:

$$D_T = \sum_{k=1}^{C} \sum_{h=1}^{C} \sum_{i=1}^{n} \sum_{\substack{j=1 \\ j \neq i}}^{n} (d+M_{ij}) A_{ij} x_{ih} x_{jk}$$

The first expression of the objective function can now be written as:

(OF2.1) $\min(D_C + D_T)$

Let us now analyze the second case, i.e. given channels speed. The flow f_{hk} from computer h to computer k (number of requests per unit time) is given by:

$$f_{hk} = \sum_{i=1}^{n} \sum_{j=1}^{n} A_{ij} x_{ih} x_{jk}$$

The capacity of channel (h,k), i.e. the maximum number of requests that can be sent from h to k, μ_{hk} is given by:

$$\mu_{hk} = T(\sum_{i=1}^{n} \sum_{j=1}^{n} \delta_{-1}(A_{ij}) x_{ih} x_{jk}) / (\sum_{i=1}^{n} \sum_{j=1}^{n} A_{ij} x_{ih} x_{jk})$$

where $\delta_{-1}(A_{ij})$ is the unitary step function:

$$\delta_{-1}(A_{ij}) = \begin{cases} 1 & \text{if } A_{ij} > 0 \\ 0 & \text{if } A_{ij} = 0 \end{cases}$$

The transmission delay on channel (h,k) is now given by:

$$W_{hk} = \frac{1}{\mu_{hk} - f_{hk}}$$

The total transmission delay D_R can be written as:

$$D_R = \sum_{k=1}^{C} \sum_{\substack{h=1 \\ h \neq k}}^{C} f_{hk} W_{hk}$$

and the objective function becomes:

(OF2.2) $\min(D_C + D_R)$

Remark that if the intercomputer flows are small with respect to the channel capacity, D_T can be considered a good approximation of D_R.

3. ALGORITHMS FOR THE TASK ASSIGNMENT PROBLEM

3.1. The combinatorial optimization problem obtained from constraints (1),(2) and (3) and objective function OF1 is a quadratic integer programming problem. It can be solved as a quadratic problem [32,37] or it can be transformed in a linear problem introducing the new binary variables y_{ijk} ($y_{ijk} = x_{ik} \cdot x_{jk}$) and the additional constraints:

$$2 \cdot y_{ijk} \leq x_{ik} + x_{jk} \tag{4}$$

$$y_{ijk} \geq x_{ik} + x_{jk} - 1 \tag{5}$$

In practice, as the problem is a maximization problem and $A_{ij} . M_{ij} \geq 0$, the (5) can be dropped without affecting the optimal solution.

In the general case the problem is NP-complete and very hard to solve [8,26,42,47,51,62];it can be efficiently solved only for particular graph topologies [2,3,19,39,42,48,50,52,60,64,66,67,73] or via heuristic algorithms [4,44,49,56,57,58].

An heuristic approach of particular interest consists in organizing the solution algorithm in as many main steps as the number of available computers (C) and, at each step, to identify, among the nodes of the computational graph, those to be assigned to a given computer such that the communications with the rest of the graph will be minimized; the nodes assigned at each step will be eliminated from the graph and are not considered in the sequel of the algorithm. The subproblem to be solved at each step is again NP-complete but efficient solution procedures can be utilized [30,40,41].

A different heuristic approach consists in introducing additional constraints in order to reduce the number of subsets of nodes to be considered as feasible clusters to assign to a computer. Examples of additional constraints leading to polynomial bounded algorithms can be found in [3,4,5,50,55,59,63].

3.2. The combinatorial optimization problems obtained from constraints (1),(2) and (3) and objective functions OF2 are are nonlinear integer programming problems and cannot be easily transformed in linear problems. Remark that for a given computer k as the left hand side of constraint (2) increase, the objective function also increase and tend to $+\infty$ for $E(x^k) \to 1^-$. Therefore all constraints (2) are satisfied with strict inequalities. But we cannot simply eliminate such constraints; in fact it is easy to verify that for $E(x^k) \to 1^+$ the objective function tend to $-\infty$ and the optimal solution is unbounded. On the other hand if

$$S \geq \sum_{i=1}^{n} A_i N_i$$ or we utilize suitable local search techniques to solve the

problem, the set of constraint (2) can be dropped.

A local search technique leading in most practical cases to good solutions (without any "a priori" guarantee) work as follows.

Let be given a feasible solution of the problem and the corresponding objective function. At each step of the algorithm take k nodes in all the possible ways and try to reallocate the k nodes in all the possible way. For each possibility calculate the corresponding value of the objective function. If a better solution is found take it as the new solution and go to the next step. Stop if no reallocation of k nodes leads to a better solution.

In practice it is not necessary to verify all the $\binom{n}{k}$ subsets of size k of the set of nodes and all the (C^k-1) reallocations of k given nodes. Several rules can be pointed out to reduce the number of possibilities to be verified and (if necessary) to avoid illegal distribution of nodes.

For k = 1 the algorithm is very efficient but the quality of the

results is in general poor, k = 3 seems to be in most cases a good compro-
mize between efficiency and quality of the solution.

3.3. Among the particular graph topologies leading to interesting algorithmic approaches, there are the tree structures. In fact many problems
in computer system design can be formalized like tree partitioning problems.

Tree partitioning (TP)

Given a weighted tree T (weights on the edges and/or on the vertices)
and a scalar B, find a partition of the vertices in clusters such that
the weight of each cluster (sum of the weights of its vertices) is not
greater than B and the connection weight (sum of the weights of all edges
with endpoints in different clusters) is minimized.

TP is simply the basic model with objective function OF1 and an underlying tree structure. TP (in decision form) is NP-complete (for $B \geq 3$)
even if T is a star or a binary tree but can be solved in pseudo-polyno-
mial time; TP is polynomial if all edge weights are equal, or if all
vertex weights are equal, or if T is a chain $(O(n^2))$; for general graphs
and $B \geq 3$ the problem remains NP-complete even if all vertex and edge
weights are 1 [42,3,60,50,34,35]. If we modify the objective function by
minimizing the number of clusters (instead of the connection weight) the
problem became polynomial and can be solved in linear time [52]; for
general graphs (even if unweighted) the problem is NP-complete.

Equipartition of trees (ET)

Given a weighted tree T and an integer m, find a partition in m non-
empty clusters such that each cluster is a tree (such a partition can be
obtained by deleting m-1 edges of T) and a norm of the m-vector v of the
differences between the clusters weight and the clusters average weight
(sum of all vertex weights divited by m) is minimized.

If T is a star, ET can be solved in polynomial time by sorting
$(O(n \log n))$ (for any norm). In the general case ET with L_∞ norm can be
solved in polynomial time. More precisely the problem of finding a

m-partition of T minimizing the maximum cluster weight can be solved by a shifting algorithm in time $O(m^3 rd(T)+m \cdot n)$ where $rd(T)$ is the number of edges in the radius of T [2]. The easiest problem of finding a m-partition of T maximizing the minimum cluster weight can be solved in time $O(m^2 rd(T)+m \cdot n)$ [66]. Remark that the same two problems for general graphs are NP-complete.

4. MULTIPLE COPIES OF NODES

If multiple copies of nodes are allowed we must introduce new sets of variables and we must modify consequently the formulation of the basic model. In the following we analyze only the model with objective function OF2.2, but all the results can be easily extended to the other simpler formulations.

In this case we must also introduce new considerations about the behaviour of the system. In fact, if two or more copies of a node exists, every time we modify the parameters or the data contained in a copy of the node we must modify in the same way also all the other copies, introducing an additional request of intercomputer flows. In order to simplify the exposition in the sequel we ignore such flows, supposing that no node modifications occur.

Under these assumptions the multiple copies problem can be formulated by introducing the variables:

y_{ij}^h = (average number of node j execution requests sent from node i to the copy h of node j).

Obviously the following constraints hold:

$$\sum_{h \in H_j} y_{ij}^h = A_{ij} \qquad \forall(i,j) \qquad (6)$$

$$y_{ij}^h \geq 0 \qquad \forall(h,i;j)$$

where H_j represents the set of node j possible copies. In order to form-
ulate correctly the model, the binary variables x_{ik} transform in:

$$x_{ik}^h = \begin{cases} 1 \text{ if the copy h of node i is located in the computer k} \\ \\ 0 \text{ else} \end{cases}$$

with the additional constraints:

$$\sum_{h \in H_i} x_{ik}^h \leq 1 \qquad\qquad \forall (i,k) \qquad\qquad\qquad (7)$$

$$\sum_{k=1}^{C} x_{ik}^h \leq 1 \qquad\qquad \forall (i,h) \qquad\qquad\qquad (8)$$

$$\sum_{k=1}^{C} \sum_{h \in H_i} x_{ik}^h \geq 1 \qquad\qquad \forall i \qquad\qquad\qquad (9)$$

the (8) and (9) replace the set of constraints (1). The (2), if needed,
will transform in:

$$\sum_{i=1}^{n} \sum_{h \in H_i} ((\sum_{j=1}^{n} y_{ji}^h) N_i / S) x_{ik}^h \leq 1 \qquad\qquad \forall k \qquad\qquad (10)$$

It is easy to see that such constraints increase of an order of magnitude
the solution algorithm. It would be therefore usefull to be able to drop
constraints (10) on the gound of what has been said in the previous section.

In the same way we can modify the quantities defining the objective
function by simply substituting y_{ij}^h to A_{ij} and considering a new problem
with $\sum_{i=1}^{n} H_i$ nodes to be located instead of n.

The overall model is quite complex and only poor heuristic solution
algorithms exist.

REFERENCES

[1]   BARNES: An algorithm for partitioning the nodes of a graph. IBM
      Waston Res. Center, RC8690, 1981.

[2]   BECKER, PERL, SCHACH: A shifting algorithm for min-max tree parti-
      tioning. J. ACM, 1982.

[3]   BERTOLAZZI, LUCERTINI, MARCHETTI: Analysis of a class of graph
      partitioning problems. RAIRO Theoretical Comp. Science, 1982.

[4]   BERTOLAZZI, LUCERTINI: Tasks assignment in a multicomputer system:
      a mathematical model. IFAC, Kyoto, 1981.

[5]   BERTZTISS: A note on segmentation of computer programs. Inf. and
      Contr., 1968.

[6]   CARLSON, NEMHAUSER: Scheduling to minimize interaction costs. Op.
      Res., 1966.

[7]   CHANDLER, DE LUTIS: A methodology of multi-criteria information
      system design. Int. Comp. Conf., 1977.

[8]   CHANDRA, WONG: Worst-case analysis of a placement algorithm related
      to storage allocation. SIAM Comp., 1975.

[9]   CHANDY: Models of distributed systems. Proc. 3rd Int. Conf. on
      VLDB, Kyoto, 1977.

[10] CHANDY, SAUER: The impact of distribution and disciplines on
      multiple processor system. Comm. ACM, 1979.

[11] CHANDY, YEH: On the design of elementary distributed systems. Comp.
      Networks, 1979.

[12] CHARNEY, PLATO: Efficient partitioning of components. ACM/IEEE Des.
      Ant. Workshop, Washington P.C., 1968.

[13] CHEN, AKOKA: Optimal design of distributed information systems.
      IEEE-TC, C-29, 1980.

[14] CHRISTOFIDES, BROOKER: The optimal partitioning of Graphs. SIAM
      J. Appl. Math., 1976.

[15]   CHU: Optimal file allocation in a computer network. In: Computer
       communication systems (ABRAMSON, KUO Eds.), Prentice Hall, 1973.

[16]   CHU, HOLLOWAY, LAN, EFE: Task allocation in distributed data pro-
       cessing Computer, 1980.

[17]   CIOFFI, COSTANTINI, DE JULIO: A new approach to the decomposition
       of sequential systems. Digital Processes, 1977.

[18]   CIOFFI, DE JULIO, LUCERTINI: Optimal decomposition of sequential
       machines via integer non-linear programming: a computational algo-
       rithm. Digital Processes, 1979.

[19]   COMEAU: A study of user program optimization in a paging system.
       ACM Symp. on Operating System Principles, Gatlinbury, 1967.

[20]   CORNUEJOLS, FISHER, NEMHAUSER: Location of Bank Accounts to opti-
       mize float: an analytic study of exact and approximate algorthms.
       Man. Sci., 1977.

[21]   DE JULIO, LUCERTINI, SACCA': Un algoritmo efficiente per la decompo
       sizione ottima di macchine sequenziali. Conf. AIRO, 1978.

[22]   DENNING: Vurtual memory. Comp. Survey, 1970.

[23]   DENNING, BUZEN: The operational analysis of queueing network models.
       Comp. Surveys, 1978.

[24]   DONATH, HOFFMAN: Lower bounds for the partitioning of Graphs.
       IBM J. of Res. and Der., 1973.

[25]   DOWDY, FOSTER: Comparative models of the file assignment problem.
       ACM Comp. Surveys, 1982.

[26]   DUTTA, KOEHLER, WHINSTON: On optimal allocation in a distributed
       processing environment. Man. Sci., 1982.

[27]   ECKHOUSE, STANKOVIC, VAN DAM: Issues in distributed processing.
       IEEE Tr. Comp., 1978.

[28]   ENSLOW: Research issues in fully distributed systems. AICA, Bari,
       1979.

[29]   FOSTER, DOWDY, AMES: File assignment in a computer network. Comp.
        Networks, 1981.

[30]   GOMORY, HU: Multi-terminal network flows. SIAM, 1961.

[31]   GORINSMTEYN: The partitioning of graphs. Eng. Cybern., 1969.

[32]   GRAVES, WHINSTON: An algorithm for the quadratic assignment pro-
        blem. Man. Sci., 1970.

[33]   GROSS, SOLAND: A branch and bound algorithm for allocation pro-
        blems in which constraint coefficients depend upon decision
        variables. Nav. Res. Log., 1969.

[34]   HADLOCK: Minimum spanning forest of bounded trees. 5th South Conf.
        on Comb., Graph Theory and Comp. Winnipeg, 1974.

[35]   HADLOCK: Finding a maximum cut of a planar graph in polynomial
        time. SIAM J. Comp., 1975.

[36]   HAESSIG, JENNY: An algorithm for allocating computational objects
        in distributed computing systems. IBM Zurich Res. Lab. RZ 1016,
        1980.

[37]   HILLIER, CONNORS: Quadratic assignment problem algorithm, and
        Location of Indivisible facilities. Man. Sci., 1966.

[38]   HOFRI, JENNY: On the allocation of processes in distributed com-
        puting systems. IBM Zurich Res. Lab. RZ 905, 1978.

[39]   HOSKEN: Optimum partitioning of the addressing structures. SIAM J.
        Comp., 1975.

[40]   HU: Integer programming and network flows. Addison-Wesley, 1970.

[41]   HU, RUSKEY: Circular cuts in a network. Math. of Op. Res., 1980.

[42]   HYAFIL, RIVEST: Graph partitioning and constructing optimal deci-
        sion trees are polynomial complete problems. IRIA Rep. 33, 1973.

[43]   IBARRA, KIM: Fast approximation algorithms for the Knapsach and
        sum of subset problems. J. ACM, 1975

[44] IBARRA, KIM: Approximation algorithms for certain scheduling pro-
      blems. Math. Op. Res., 1978.

[45] JENNY, KUMMERLE: Distributed processing within an integrated
      circuit/packet-switching node. IEEE Tr. Commun., 1976.

[46] JENSEN: Optimal network partitioning. Op. R ls., 1970.

[47] JOHNSON: Approximation algorithms for combinatorial problems.
      J. Comp. and Syst. Sci., 1974.

[48] KERNIGHAN: Some graph partitioning problems related to program
      segmentation. Ph. D. Th., Princeton, 1969.

[49] KERNIGHAN, LIN: An Efficient Heuristic Procedure for Partitioning
      Graphs. Bell System Tech. J., 1970.

[50] KERNIGHAN: Optimal sequential partitions of graphs. J. ACM, 1971.

[51] KOONTZ, NORENDRA, FUKUNAGA: A branch and bound clustering algorithm.
      IEEE Tr. Comp., 1975.

[52] KUNDU, MISRA: A linear tree partitioning algorithm. SIAM J. Comp.,
      1977.

[53] LAWLER: Electrical assemblies with a minimum number of intercon-
      nections. IKE Tr. Elec. Comp., 1962.

[54] LAWLER, LEVITT, TURNER: Module clustering to minimize delay in
      digital networks. IEEE Tr. Comp., 1969.

[55] LAWLER: Cutsets and partitions of Hypergraphs Networks, 1973.

[56] LAWLER: Fast approximation algorithms for Knapsack problems. Math.
      Op. Res., 1979.

[57] LIPTON, TARJAN: A separator theorem for planar graphs. Conf. on
      Theor. Comp. Sci., Waterloo, 1977.

[58] LIPTON, TARJAN: Applications of a planar separator theorem. 18th
      FOCS. Long Beach, 1977.

[59] LUCCIO, SAMI: On the decomposition of networks in minimally inter-connected subnetworks. IEEE Tr. Circ. theory, 1969.

[60] LUKES: Efficient algorithm for the partitioning of trees. IBM J. Res. Der., 1974.

[61] MAHLOUD, RIORDAN: Optimal allocation of resources in Distributed Information Networks. ACM Tr. Databases Syst., 1976.

[62] MARSTEN: An algorithm for large set partitioning problems. Man. Sci., 1974.

[63] MENDELSON, PLISKIN, YECHIALI: Optimal storage allocation for serial files. Comm. ACM, 1979.

[64] MISRA, TARJAN: Optimal chain partitions of trees. Inf. Proc. Letters, 1975.

[65] MOKGAN, LEVIN: Optimal program and data locations in computer networks. Comm. ACM, 1977.

[66] PERL, SCHACH: Max-min tree partitioning. J. ACM, 1981.

[67] PERL, SMILOACH: Efficient optimization of monotonic function on trees. CAAP 81, Lect. Notes in Cong. Sci., 1981.

[68] RAO, STONE, HU: Assignment of tasks in a distributed processor system with limited memory. IEEE Tr. Comp., 1979.

[69] RUSSO, ODEN, WOLFF: A Heuristic procedure for the partitioning and mapping of computer logic blocks to modules. IEEE Tr. Comp., 1971.

[70] SACCA', WIEDERHOLD: Database Partitioning in a cluster of pro-cessors. VLDB Conf., 1983.

[71] SAHNI: Approximative algorithms for the 0/1 Knapsack problem. J. ACM, 1975.

[72] SAHNI: General techniques for combinatorial approximation. Op. Res., 1977.

[73] SCHRADER: Approximations to clustering and subgraph problems on trees. Okonometric and Op. Res. Inst. Rep. 81202-OR, Bonn, 1981.

[74] SIMEONE: An asymptotically exact polynomial algorithm for equi-partition problems. IAC n. 153, Roma, 1978.

[75] STONE: Multiprocessor scheduling with the aid of network flow algorithms. IEEE Tr. Soft. Eng., 1977.

[76] TARJAN: A hierarchical clustering algorithm using strong components. Int. Proc. Letters, 1982.

[77] TARJAN: An improved algorithm for hierarchical clustering using strong components. Int. Proc. Letters, 1983.

[78] WONG, COPPERSMITH: A combinatorial problem related to multimodule memory organization. J. ACM, 1974.

APPROXIMATION ALGORITHMS FOR BIN-PACKING — AN UPDATED SURVEY

E.G. Coffman, Jr.
M.R. Garey
D.S. Johnson

Bell Laboratories
Murray Hill, New Jersey 07974

1. Introduction

This paper updates a survey [53] written about 3 years ago. All of the results mentioned there are covered here as well. However, as a major justification for this second edition we shall be presenting many new results, some of which represent important advances. As a measure of the impressive amount of research in just 3 years, the present reference list more than doubles the list in [53].

Characteristic of bin-packing applications is the necessity to pack or fit a collection of objects into well-defined regions so that they do not overlap. From an engineering point of view the problem is normally one of making efficient use of time and/or space. A basic mathematical model is defined in the classical one-dimensional bin packing problem: We are given a positive integer bin capacity C and a set or *list* of items $L = (p_1, p_2, ..., p_n)$, each item p_i having an integer size $s(p_i)$ satisfying $0 \leqslant s(p_i) \leqslant C$. What is the smallest integer m such that there is a partition $L = B_1 \cup B_2 \cup ... \cup B_m$ satisfying $\sum_{p_i \in B_j} s(p_i) \leqslant C$, $1 \leqslant j \leqslant m$? We usually think of each set B_j as being the contents of a *bin* of capacity C, and view ourselves as attempting to minimize the number of bins needed for a packing of L.

By this choice of terms the obvious interpretation of bin-packing corresponds to problems of storage. However, the variety of other interpretations that can be placed on the parameters and terminology accounts for the fundamental importance of the problem. Packing trucks with a given weight limit and assigning commercials to station breaks on television [11] illustrate this variety in the real world. A commonly cited, general example is the following cutting-stock problem. Material such as cable, lumber, pipes, tapes, etc. is supplied in a standard length, C. Demands for pieces of the material are for arbitrary lengths not exceeding C. The problem is to use the minimum number of standard lengths in accommodating a given list of required pieces.

Problems in which time is the dimension (resource) being partitioned are represented by the following scheduling problem: We are given a collection of identical processors on which a set of independent tasks with known execution times are to be executed. The problem is to determine the least number of processors that must be used in order that all tasks be completed by some given deadline. Here, the processors are bins, the deadline is the common bin capacity and the elements of L are the task execution times.

This problem establishes the connection between bin-packing and combinatorial scheduling theory. Note in particular the close relationship between this problem and the multiprocessor scheduling problem, i.e. the problem of minimizing makespan on parallel processors. (The makespan or length of a schedule is simply the latest task finishing time.) In bin-packing terms it corresponds to the bin design problem: Given L and a fixed set of m bins, what is the least capacity C such that L can be packed in m bins of this capacity? Historically, research on this problem was carried out exclusively under the heading of scheduling theory. These results are included here because *both* the classical and capacity minimization problems have significance in scheduling and storage applications, and because they are based on the same mathematical structure, i.e. they differ only in the descriptor held fixed and the one chosen as the objective function.

It is consistent with the effort invested in these problems that efficiently computing optimal solutions has proved to be quite difficult. In fact, the bin-packing problem, or more precisely the decision problem "Given C, L, and an integer bound K, can L be packed into K or fewer bins of capacity C?" is NP-complete. A similar statement holds for the decision problem corresponding to multiprocessor scheduling. By the theory elaborated in [52,71,78], this means that it is unlikely that efficient, (i.e., polynomial time) optimization algorithms can be found for these problems. Thus researchers have turned to the study of approximation algorithms, that is, algorithms which, although not guaranteed to find an optimal solution for every instance, attempt to find *near*-optimal solutions. The analysis of approximation algorithms is the dominant topic in the remainder of this paper. It is primarily this theme that has determined the literature we have chosen to survey. This theme will be further narrowed to those relatively simple but effective algorithms which have been successfully analyzed for measures of worst-case or average-case performance.

Along with closely related partitioning problems, bin-packing and multiprocessor scheduling have played an important role in applications of complexity theory [52]. They also hold a special place in the history of approximation algorithms. It was in these contexts that the first work was done in proving that fast approximation algorithms could actually *guarantee* near-optimal solutions. The early work of Graham [60,62] on multiprocessor scheduling inaugurated this approach, and the early work in bin packing (most notably [72]) served to popularize and extend the methodology.

The scope of applications has been widened considerably by the study of a number of variants of the basic problem. Approximation algorithms have been designed and analyzed for the following four basic modifications: (1) Packings in which bounds are placed in advance on the number of items that can be packed in a bin, (2) Packings in which a partial order is associated with the set of items to be packed and constrains the ways in which items may be packed. (3) Packings in which restrictions are placed on the items that may be packed in the same bin, and (4) Packings in which items may enter and leave the packing dynamically. These variants will be covered in Section 3 following the survey of results in Section 2 for the classical problem.

Multiprocessor scheduling results will be described in Section 4. In addition to this problem, there have been several others based on optimizing objective functions other than the number of bins. In Section 5 we shall consider such objective functions as the number of items packed and the sum of the squares of the bin levels, where bin level refers to the total size of the items in a bin.

With the techniques that had developed for the one-dimensional problems it was natural that efforts eventually turn to higher dimensions. In Section 6 we discuss the results on vector packing where items sizes and bin capacity are assumed to be d-dimensional vectors. Such problems model scheduling applications in which jobs must use several different resources during their execution.

In Section 7 we survey the large and growing literature on two dimensional packing. Once again the obvious industrial applications in stock cutting have been an important stimulus to this research. Further motivation has been provided by advances in VLSI technology in which layouts on chips pose a number of important combinatorial packing problems. The focus of this survey on the analysis of approximation algorithms essentially limits us to the research on packing rectangular figures into two dimensional "bins" or strips.

We shall be covering numerous improvements to the early results in bin-packing that were not mentioned in [53]. This includes the discovery of polynomial approximation schemes for one-dimensional bin packing, and the many new results on the probabilistic analysis of packing algorithms. In the final section we shall mention a few of the many open problems still outstanding.

2. The Classical Bin-Packing Problem

We begin by describing three basic algorithms for the problem as defined in the preceding section. The first, and simplest, is NEXT-FIT: We process the items in L in turn, starting with p_1, which is placed in bin B_1. Suppose that p_i is now to be packed, and let B_j be the highest indexed non-empty bin. If p_i will fit in B_j (the level of B_j does not exceed $C - s(p_i)$), then put p_i in bin B_j. Otherwise, start a new bin (bin B_{j+1}) by putting p_i into it.

This is clearly a fast algorithm (linear time). Moreover, it is not difficult to show that, if $NF(L)$ is the number of bins used in the NEXT FIT packing of list L and $OPT(L)$ is the number of bins required in an optimal packing, then for all lists L, $NF(L) \leqslant 2 \cdot OPT(L)$. This is the best bound of this sort we can prove for NEXT FIT, since the examples shown in Figure 1 indicate that there are lists L with $NF(L) \geqslant 2 \cdot OPT(L) - 1$.

$$L = (\tfrac{1}{2}, \tfrac{1}{2N}, \tfrac{1}{2}, \tfrac{1}{2N}, \ldots, \tfrac{1}{2}, \tfrac{1}{2N})$$

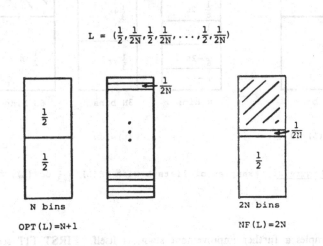

$$OPT(L) = N+1 \qquad\qquad NF(L) = 2N$$

Figure 1. Examples of lists L with NF(L) = 2·OPT(L) - 1.

To improve on this bound we need a new algorithm. One defect of NEXT FIT seems to be that it only tries to put p_i in one bin before it resorts to starting a new bin. This suggests that the following FIRST FIT algorithm might be an improvement: When packing p_i, put it in the lowest indexed bin into which it will fit (starting a new bin only if p_i will not fit into any non-empty bin). It can be shown (though the proof [49,72] is more difficult) that for all lists L,

$FF(L) \leqslant (17/10) \cdot OPT(L) + 1$ [72] and, again, this is the best ratio possible, since there are lists L with arbitrarily large values of $OPT(L)$ such that $FF(L) \geqslant (17/10) \cdot OPT(L) - 8$ [72]. These lists are too complicated to illustrate here, but Figure 2 shows examples that approach a ratio of $5/3 = 1.6666....$

$$L = (\tfrac{1}{6} - 2\epsilon, \ldots, \tfrac{1}{6} - 2\epsilon, \tfrac{1}{3} + \epsilon, \ldots, \tfrac{1}{3} + \epsilon, \tfrac{1}{2} + \epsilon, \ldots, \tfrac{1}{2} + \epsilon)$$

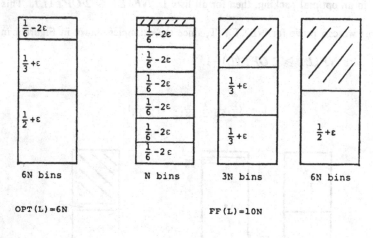

Figure 2. Examples of lists L with $FF(L) = \tfrac{5}{3} \cdot OPT(L)$.

From these examples a further improvement suggests itself. FIRST FIT seems to perform poorly when the large items occur at the end of the list. The algorithm FIRST FIT DECREASING seeks to avoid this effect by first ordering the items so that $s(p_1) \geqslant s(p_2) \geqslant \ldots \geqslant s(p_n)$, and then applying FIRST FIT to the reordered list. For this algorithm it can be shown (with considerable difficulty [4,69,72]) that for all lists L,

$FFD(L) \leqslant \dfrac{11}{9} \cdot OPT(L) + 4$ and, once more, this is the best ratio possible, as illustrated in Figure 3.

$$L = (\tfrac{1}{2}+\varepsilon,\ldots,\tfrac{1}{2}+\varepsilon,\tfrac{1}{4}+2\varepsilon,\ldots,\tfrac{1}{4}+\varepsilon,\ldots,\tfrac{1}{4}+\varepsilon,\tfrac{1}{4}-2\varepsilon,\tfrac{1}{4}-2\varepsilon)$$

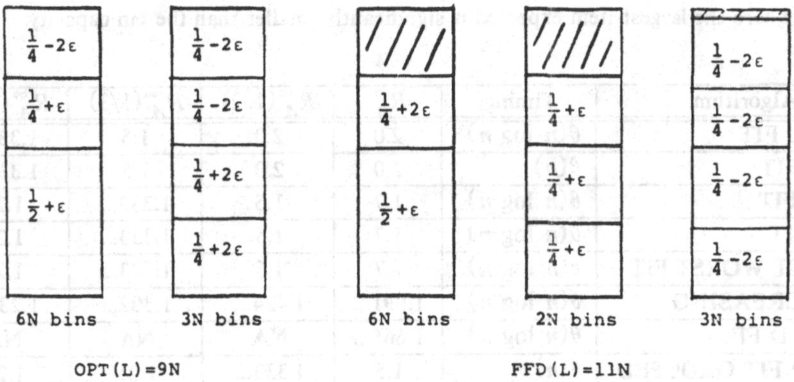

OPT(L)=9N   FFD(L)=11N

Figure 3. Examples of lists L with FFD(L)=$\frac{11}{9}$OPT(L).

Let us formalize the type of worst case analysis we have been discussing. If A is an algorithm and $A(L)$ is the number of bins used by that algorithm for list L, define $R_A(L) \equiv A(L)/OPT(L)$. The *absolute performance ratio* R_A for algorithm A is given by

$$R_A \equiv \inf\{r \geqslant 1: R_A(L) \leqslant r \text{ for all lists } L\}.$$

The *asymptotic performance ratio* R_A^{∞} for A is given by

$$R_A^{\infty} \equiv \inf\{r \geqslant 1: \text{ for some } N > 0,\, R_A(L) \leqslant r \text{ for all } L \text{ with } OPT(L) \geqslant N\}.$$

The above results can now be summarized by saying that $R_{NF}^{\infty} = 2$, $R_{FF}^{\infty} = 17/10$, and $R_{FFD}^{\infty} = 11/9$. Notice that R_A need not equal R_A^{∞}. Although $R_{FFD}^{\infty} = 11/9$, it is easy to give lists L for which $OPT(L) = 2$ and $FFD(L) = 3$, so that $R_{FFD} \geqslant 3/2$. The asymptotic ratios seem to be a more reasonable measure of performance for the basic bin packing problem, but absolute ratios do come up in some of the work on related problems that we shall be discussing later.

Table 1 highlights the early results that were obtained for several other algorithms, along with those just described. The quantity $R_A^\infty(t)$, $0 < t \leqslant 1$ is the asymptotic worst case ratio for algorithm A on lists all of whose items have size bounded by $t \cdot C$. This measure is of interest in applications where the largest item expected is significantly smaller than the bin capacity.

Algorithm	Timing	R_A^∞	$R_A^\infty(1/2)$	$R_A^\infty(1/3)$	$R_A^\infty(1/4)$
WORST FIT	$\theta(n \log n)$	2.0	2.0	1.5	1.333...
NEXT FIT	$\theta(n)$	2.0	2.0	1.5	1.333...
FIRST FIT	$\theta(n \log n)$	1.7	1.5	1.333...	1.25
BEST FIT	$\theta(n \log n)$	1.7	1.5	1.333...	1.25
ALMOST WORST FIT	$\theta(n \log n)$	1.7	1.5	1.333...	1.25
NF DECREASING	$\theta(n \log n)$	1.691...	1.424...	1.302...	1.234...
REVISED FF	$\theta(n \log n)$	1.666...	NA	NA	NA
GROUP FIT GROUPED	$\theta(n)$	1.5	1.333...	1.25	1.20
FF GROUPED	$\theta(n \log n)$	1.333...	1.333...	1.25	1.20
ITERATED LFD	$\theta(n \log^2 n)$	1.333...	NA	NA	NA
FF DECREASING	$\theta(n \log n)$	1.222...	1.183...	1.183...	1.15
BF DECREASING	$\theta(n \log n)$	1.222...	1.183...	1.183...	1.15
MODIFIED FFD	$\theta(n \log n)$	1.183...	1.183...	1.183...	1.15

Table 1. Asymptotic worst case bounds for bin packing algorithms.

The algorithms REVISED FIRST FIT and MODIFIED FIRST FIT DECREASING are recent developments which we shall be discussing in detail shortly. Most of the other results in the table were already known by 1973 [69,70]. The algorithm BEST FIT (BF) is like FIRST FIT, except that p_i is placed in the bin into which it will fit with the smallest gap left over (with ties broken in favor of the lowest indexed bin) [72]. WORST FIT (WF) [69,70] places p_i in the non-empty bin with the biggest gap (ties broken in the same way), starting a new bin if this biggest gap is not big enough. ALMOST WORST FIT (AWF) [69,70] tries the second largest gap first, and then proceeds as does WORST FIT — surprisingly, this makes a difference. The analysis of NEXT FIT DECREASING was done by Baker and Coffman [7]. GROUP FIT GROUPED (GFG) [69,70] uses "implicit rounding" to discretize the ranges of item sizes and bin levels, thus avoiding the sorting implicit in the FFD algorithm which it mimics. It also attains a linear running time, while paying only a partial penalty in worst case behavior. FIRST FIT GROUPED (FFG) [69,70] is a hybrid algorithm, included mainly because it yields a different value of R_A^∞. The algorithm ITERATED LOWEST FIT DECREASING, which attains the same value, but more slowly, is due to Krause, Shen, and Schwetman [85], and will be discussed in more detail in the next section.

A variety of other results were also obtained during the early 1970's. The precise values of $R_A^\infty(t)$ as a function of t were obtained for many of the algorithms [69,70,72]. Except for the algorithms WORST FIT and NEXT FIT, which yield the continuous function $R_A^\infty(t) = 1+t(1-t)$, these tend to be step functions determined by $\lfloor 1/t \rfloor$. In [69] the asymptotic worst case behavior of FIRST FIT was completely determined for the case when all item sizes lie in a specified interval, as a function of the interval. The algorithms NEXT-k FIT, $k \geqslant 1$, which resemble NEXT FIT except that p_i is placed in a new bin only if it will not fit in any of the last k non-empty bins (NEXT-1 FIT is the same as NEXT FIT), were studied in [69,70]. These papers also analyzed what might be considered "non-deterministic" bin packing algorithms: ANY FIT (AF), which can place p_i anywhere, except that it can never put it in a new bin unless it won't fit in any of the already non-empty bins, and ALMOST ANY FIT (AAF), which in addition can never put p_i in a bin whose gap is larger than that of all other bins unless that is the only place it fits. The results for ALMOST ANY FIT are the same as those for FIRST, BEST, and ALMOST WORST FIT (all of which obey the AAF assumptions), while the results for ANY FIT are the same as those for WORST FIT, which essentially makes the worst choices allowed under the AF assumptions. ANY FIT DECREASING and all DECREASING algorithms obeying the ANY FIT ground rules seem to obey the same bounds as FFD, although the best that has been proved for any such algorithms (other than FFD and BFD) is that $R_A^\infty \leqslant 5/4 = 1.25$ [69,70].

Another special class of algorithms that has received attention consists of the "on-line" algorithms. An on-line algorithm is one which, like NEXT FIT or FIRST FIT, assigns items to bins in exactly the order they are given in the original list, without using any knowledge about subsequent items in the list. FIRST FIT DECREASING, for example, is not an on-line algorithm, since it first re-orders the list. On-line algorithms may be the only ones that can be used in certain situations, where the items to be packed are arriving in a sequence according to some physical process and must be assigned to a bin as soon as they arrive. Thus, although it is known that "off-line" algorithms such as FFD can do much better than FIRST FIT, it is of interest to determine the best worst-case performance that any on-line algorithm can have. On the basis of a clever analysis of the worst case examples for FIRST FIT, Yao [109] was able to devise a new algorithm, REVISED FIRST FIT (RFF), with $R_{RFF}^{\infty} = 5/3 = 1.6666$, which is to be compared with $R_{FF}^{\infty} = 1.7$. Even more significantly, he was able to show that for *any* on-line algorithm A, we must have $R_A^{\infty} \geqslant 1.5$. In subsequent work Brown [12] and Liang [91] independently extended Yao's lower bound results, improving the lower bound to 1.536. In addition Brown designed a further revision of FIRST FIT, whose asymptotic worst case ratio is better than 1.64 [14].

Galambos and Turan [46] quite recently considered the lower bound question for on-line algorithms when the list is assumed to be in non-increasing order. They showed that any such algorithm must have a worst-case bound that is at least 10/9.

In [109] a slight improvement to FFD was also found. However, more significant improvements with an $O(n \log n)$ running time not much worse than that of FFD were discovered by Friesen and Langston [45] and by Garey and Johnson [54]. The first of these employs a hybrid algorithm: Both FFD and an algorithm called BEST TWO FIT are run on the input; the output is taken as the better of the two packings produced. Friesen and Langston showed that, for any list, the *average* of the number of bins required by the two component algorithms can not exceed $6/5 = 1.2$, thus guaranteeing a weak upper bound on the minimum.

Garey and Johnson devised the MODIFIED FIRST FIT DECREASING algorithm for which the tight asymptotic bound, $R_{MFFD}^{\infty} = 71/60 = 1.18333...$, was proved. The algorithm is based on a careful analysis of the 11/9 examples for FFD and what causes FFD to mispack them. It

proceeds as follows: Partition the input list L into three sublists $L_A = \{p_i: s(p_i) \in (\frac{C}{3}, C]\}$,

$L_D = \{p_i: s(p_i) \in (\frac{11C}{71}, \frac{C}{3}]\}$ and $L_X = \{p_i: s(p_i) \in (0, \frac{11C}{71}]\}$. The first step is to pack the

sublist L_A using *FFD*. In the resulting packing, call a bin containing only a single item from L_A

an "A-bin." Then pack as much of L_D into A-bins as possible using the following loop:

1. Let bin B_j be the A-bin with the currently largest gap. If the two smallest unpacked items in

    L_D will not fit together in B_j, exit from the loop.

2. Let p_i be the smallest unpacked item in L_D, and place p_i in B_j.

3. Let p_k be the *largest* unpacked item in L_D that will now fit in B_j, and place p_k in B_j. Go to

    1.

The assignment of items to bins is then completed by combining the unpacked portion of L_D with

L_X and adding all these remaining items to the packing using *FFD*.

It should be noted that the proofs of the results for the more effective algorithms are

characteristically long and intricate "weighting function" arguments. This proof technique

originated with the analysis of FIRST-FIT [72], and plays a central role in the theory. A tutorial

discussion of the use of weighting functions can be found in [22].

A number of the detailed proofs have been so long as to preclude their full publication in

technical journals (e.g. there are results in [33,72,86] whose proofs span 100 pages). Baker [4] has

illustrated, however, that significantly shorter proofs may be possible; Baker gives a proof of the

basic 11/9 theorem which is about 1/3 the length of the original.

Other Item Constraints — As we have seen, when all piece sizes are known to be sufficiently small

compared to C, the performance of approximation rules can improve substantially. Similar improvements can be expected from other restrictions that occur frequently in practice. For example, suppose all item sizes are of the form $C(\frac{1}{k})^j$, $j \geqslant 0$, for some fixed positive integer k. Then it is not difficult to show that $FFD(L) = OPT(L)$. As shown in [28] similar results hold for many of the approximation algorithms designed for the problems surveyed in the remaining sections of this paper. Note that power-of-two item sizes occur in important computer applications; by the nature of binary computers, if the sizes of records (files, pages, etc.) are constrained to be powers of two, algorithms for maintaining and allocating storage have much more efficient implementations.

As another illustration suppose the number of different item sizes is fixed, and therefore the number of bin types (i.e. the number of possible item-size configurations that fit into a bin) is finite. In this case, the work of Gilmore and Gomory [55,56] in the early 60's can be applied. (The importance of this work will emerge again shortly in our discussion of approximation schemes.) They were able to show that the linear programming relaxation of the problem, although still quite large (it has a variable for each bin type), can be solved using special techniques. An actual packing is then constructed by "rounding up" the solution values. In terms of worst case analysis, this algorithm will have $R_A^\infty = 1$ for any fixed number of item sizes, since it can yield at most one excess bin for each possible bin type (a much larger, but still fixed number, independent of the number of items). We note in passing that when the number of item sizes is fixed, we actually can find *optimal* solutions in polynomial time, although the degree of the polynomial can be astronomical. Gilmore and Gomory's contribution is in obtaining *almost* optimal solutions with much less work.

Approximation Schemes — The prospects for improved approximation algorithms came to be much better understood as the result of two major results of the past three years. The first was the discovery by Fernandez de la Vega and Lueker [40] that for every $\epsilon > 0$ there is a linear-time algorithm, $A[\epsilon]$, with $R_{A[\epsilon]}^\infty \leqslant 1+\epsilon$. Algorithms of this type had long been known for problems such as the knapsack problem, where performance is measured by absolute, rather than asymptotic worst-case ratios. A set of such algorithms $\{A[\epsilon]: \epsilon > 0\}$ has been termed an *approximation scheme*. It was shown in [40] that techniques from the knapsack approximation scheme could in fact be used in designing bin-packing algorithms satisfying

$$A[\epsilon](L) \leqslant (1+\epsilon)OPT(L)+(1/\epsilon)^2 ,$$

and having a running time linear in the length of L for fixed ϵ. The central idea in the proof is the reduction of the original bin-packing problem to one in which the number of possible item configurations in a bin is bounded. As in the Gilmore and Gomory [55] approach mentioned earlier the algorithm is formulated as a solution to a linear program.

Subsequently, Karmarkar and Karp [74] eliminated a shortcoming of the above result, viz. the fact that the running time of $A[\epsilon]$ is exponential in $(1/\epsilon)^2$. Using the Fernandez de la Vega and Lueker results and an impressive array of techniques from mathematical programming and complexity theory they devised a "fully polynomial" approximation scheme, i.e. one for which the running time is a polynomial in both $1/\epsilon$ and the length of L, and the additive constant is also a polynomial in $1/\epsilon$.

An interesting corollary to these results is that there exist polynomial time approximation algorithms with $R_A^\infty = 1$. One need only choose ϵ as an appropriate function of the given instance. Unfortunately, the actual guarantee provided by these algorithms is not $OPT(L)+K$ for some constant, K, but $OPT(L)+f(OPT(L))$, where f is a slowly growing function. The best such function comes from Karmarkar and Karp's analysis, and is $O(\log^2(OPT(L)))$.

At present the above results are mainly theoretical in their significance, because the coefficients hidden in the term "polynomial time" are too large for practical purposes. However, in principle at least, the search for better approximation algorithms can now take a different tack: Instead of trying to improve old bounds without great sacrifices in running time, we can try to improve on old running times without great sacrifices in the bound.

Average Case Analysis — As might be expected from the greater difficulty in calculating probabilistic measures, the known results for the average case are less sharp and less general than those for the worst-case. However, as we shall see, the field is very active and a number of significant advances can be cited.

Before getting into the analytical approach let us consider what has been learned from Monte Carlo simulations. The most extensive experiments appear to be those in [69], and subsequently those in [96]. Our illustrations will be drawn from [69]. Since the results in [96] measure percentage of waste per bin rather than number of excess bins, they are not readily comparable with our worst case results. A summary of some of the results is shown in Table 2.

Algorithm	UNIFORM (0,1.0)	UNIFORM (0,0.5)	UNIFORM (0,0.25)
NEXT FIT	31.1 [100.]	18.8 [100.]	7.4 [50.0]
NEXT-2 FIT	21.9 [85.0]	8.5 [50.0]	2.2 [25.0]
ALMOST WORST FIT	10.4 [70.0]	4.8 [50.0]	1.4 [25.0]
FIRST FIT	7.0 [70.0]	2.2 [50.0]	0.6 [25.0]
BEST FIT	5.6 [70.0]	2.2 [50.0]	0.5 [25.0]
GROUP FIT GROUPED	2.1 [50.0]	0.4 [33.3]	0.3 [20.0]
AWF DECREASING	2.0 [22.2]	0.2 [18.3]	0.2 [15.0]
FF DECREASING	1.9 [22.2]	0.1 [18.3]	0.2 [15.0]
BF DECREASING	1.9 [22.2]	0.1 [18.3]	0.2 [15.0]

Table 2. Percentages of excess bins required on the average in bin-packings of 25 200-item lists with item sizes uniformly distributed within the stated ranges. [Percentage of excess in worst examples known are given in brackets.]

We should note that the ratios given are not strictly speaking averages of $R_A(L)$, since the value of $OPT(L)$ could not be determined (its computation being an NP-complete problem). Instead these values are for the ratio of $A(L)$ to the sum of the item sizes. As we shall see later, however, there is strong evidence to support the claim that this approximation loses very little when L is large.

The interesting fact from these simulations is that the average behavior, although much better than the worst case behavior, still ranks the algorithms in the same relative order. Results for an approximation to a normal distribution, and for a distribution obtained by partitioning a set of items of size C into a random number of items, are slightly worse, but reflect the same trends [37]. It should not be expected, however, that average case ranking will always reflect worst case ranking. In particular, certain of the new algorithms specifically designed for improved worst case behavior (although possibly not MFFD or the hybrid algorithm in [45]) may be comparatively bad on the average.

Because of the restriction to 200-item list the results in Table 2 are not indicative of asymptotic behavior. Extensive simulation studies of asymptotic performance are currently in progress by Jon Bentley and Catherine McGeoch at Bell Laboratories. An interesting sample of their early results is that the expected performance of FF is $n/2 + H(n^{0.8})$ for C normalized to 1 and for n items with sizes uniformly distributed over [0,1].

The first mathematical results appeared in an approximate analysis by Shapiro [102]. The approximation was based on the exponential distribution and estimated the expected value, given $NF(L)$, of $OPT(L)$. He concluded that as $NF(L)$ approaches infinity, $R_{NF}(L)$ approaches 1 plus the average item size, when that average is $C/5$ or less.

The first exact results were obtained by Coffman, So, Hofri and Yao [32] from an analysis of a Markov process defined on the bin levels. For item sizes uniformly distributed between O and C, it was shown that $E(NF(L))$ is bounded by $(4/3)E(OPT(L))+4$. It was also shown that convergence to the stationary NEXT FIT bin-level distribution was exponentially fast. Hofri [68] and Ong, Magazine and Wee [97] have recently strengthened these results by working out the second moment of $NF(L)$, and Hofri has derived an approximate probability generating function valid for general item size distributions.

Karmarkar [73] subsequently extended the NF results to cover uniform distributions over any interval $(0, tC]$. Closed forms were obtained for $\frac{1}{2} \leqslant t \leqslant 1$, and gave results within about .5% of the empirical results in [96]. It is perhaps natural to expect that the stationary, expected bin-level would increase monotonically with decreases in the maximum piece size, t. However, as observed in the data of [96] and confirmed analytically in [73], this is not the case for NEXT FIT. In fact, packing efficiency was found to be least at $t = 0.841$, which agrees with empirically observed behavior.

Results on algorithms other than NEXT FIT have also been obtained for uniformly distributed item sizes. In [42], Frederickson gives a rigorous proof of the rather intuitive fact that, for this particular distribution, the ratio $E(FFD(L))/E(OPT(L))$ approaches 1 as $n \rightarrow \infty$, where n is the number of items. (Note that this ratio is not the same as $E(FFD(L)/OPT(L))$.) Frederickson achieves this result by analyzing a different algorithm which produces packings with a much simpler structure. Basically, working inward from both ends of an ordered list of the pieces, the algorithm attempts to pair large pieces with small ones. The analysis shows that the algorithm requires $n/2 + O(n^{2/3})$ bins, on the average, to pack a list of n pieces with $C = 1$.

Early results on the asymptotic properties of optimal packings were obtained by Loulou [93]. He proved that if the n item sizes are independent and identical random variables with the density function f, then the limit in probability, $\bar{R}$, of the ratio of $E(OPT(L))$ to the expected total of the item sizes divided by C is 1 as $n \rightarrow \infty$, if f is symmetric or if f is positive and decreasing over $[O, C]$. Karmarkar [65] recently showed that this asymptotic result also holds if f is decreasing over $[O, tC]$, $0 < t < 1$, and 0 elsewhere.

Results of this type have important consequences for the analysis of approximation algorithms. They suggest that for reasonably large lists, the approximation of $OPT(L)$ by the sum of the item sizes divided by C is indeed a very good one for large classes of item size distributions.

Lueker [95] studied the same question specialized as follows: If f is uniform over $[a,b]$, $b > a$, for what values of a and b does $\bar{R} = 1$? For $a = 0$ or $a+b = 1$ it was known that $\bar{R} = 1$, and for points in the region $b > 1-a$ it was known that $\bar{R} > 1$. Lueker identified a substantial subset of the region $0 < a < b < 1-a$ where $\bar{R} > 1$, and cited simulation results which suggest that $\bar{R} = 1$ elsewhere in this region.

Knödel [82] and Lueker [94] have extended Frederickson's results for the uniform distribution to other approximation algorithms. With a slight modification of Frederickson's algorithm $(C=1)$ Lueker showed that the expected number of bins used can be tightened to $n/2 + O(n^{1/2})$, a result that also applies to Knödel's algorithm. Within a multiplicative constant this is the best possible performance, since Lueker also showed that an *optimal* algorithm must use $n/2 + \Theta(n^{1/2})$ bins on the average. In [76] Karmarkar, Karp, Lueker and Murgolo generalized these results to any probability density symmetric about $C/2$ or positive and decreasing over $[0,C]$, thus strengthening Loulou's earlier result.

Hoffmann [66] and more recently Lee and Lee [90] have considered more deeply the expected performance of *on-line* algorithms. They have designed algorithms based on "reservation" techniques whereby bins are dedicated to particular configurations of items. In [66] these configurations are based on those of Frederickson's pairing algorithm. Hoffmann has shown that his on-line algorithm retains asymptotic optimality (in the earlier expected value sense), but at the expense of a poor worst-case performance.

Lee and Lee develop an algorithm with a good worst-case performance and with a good, though not asymptotically optimal, expected performance that is relatively easy to analyze. Their on-line HARMONIC algorithm, H, packs items so that the sizes of all items in any given bin are in the same interval of the set $\{(\frac{C}{2},C], (\frac{C}{3},\frac{C}{2}],...,(0,\frac{C}{M}]\}$, where $M \geqslant 2$ is a parameter of the algorithm. They show that $R_H^\infty(M) \leqslant 1.692$ for all $M \geqslant 12$, and that

$\lim\limits_{M \to \infty} R_H^\infty(M) = 1.691\ldots$, thus improving on both NEXT FIT and FIRST FIT. The latter

bound,

$$1.691\ldots = 1 + \frac{1}{2} + \frac{1}{2\times 3} + \frac{1}{6\times 7} + \frac{1}{42\times 43} + \ldots ,$$

is the same as the one given for NF DECREASING in Table 1. (A worst-case analysis of NF

DECREASING [7] also entails the analysis of HARMONIC packings.)

The average-case performance of HARMONIC is analyzed for several item size distributions.

For the uniform distribution over $[0,C]$ they show that $E(H(L))/E(OPT(L))$ is asymptotically

no greater than 1.29 for all $M > 12$, which is to be compared with the corresponding 4/3 result

for NEXT FIT.

3. Bin-Packing Variants

In this section we survey results for variants on the classical one-dimensional bin packing

problem in which the goal is still to minimize the number of bins used.

Constraints on the Number per Bin — This modification was considered by Krause, Shen, and

Schwetman [85] as a model for multiprocessor scheduling under a single resource constraint when

the number k of processors is fixed. In this case the items represent tasks to be executed, with the

size of an item being the amount of the resource it requires (out of a total quantity of C). If we

assume that all tasks have the same unit-length execution time, then a schedule corresponds to an

assignment of tasks to integral starting times, such that at no time are there more than k tasks

being executed or is there more than C of the resource being used. The objective is to minimize the

latest starting time. This corresponds to bin packing where the bins represent starting times and can

contain at most k items.

Krause et al. analyze three algorithms for this problem. The first two are just FIRST FIT and

FIRST FIT DECREASING, suitably modified to account for the bound on the number of items per bin. The results are simply stated:

$$\frac{27}{10} - \left\lceil \frac{37}{10k} \right\rceil \leq R_{FF}^{\infty} \leq \frac{27}{10} - \frac{24}{10k}; \quad R_{FFD}^{\infty} = 2 - \frac{2}{k}$$

Note that as $k \rightarrow \infty$, these bounds remain substantially worse than the corresponding bounds when the number of items per bin is not restricted (27/10 versus 17/10 and 2 versus 11/9). Thus the very existence of a limit, and not just its size, can have a substantial effect on the worst case behavior of the algorithms.

The third algorithm studied was alluded to in the previous section. ITERATED LOWEST FIT DECREASING uses a technique we shall be meeting again in the next section, so we shall describe it in detail. We first put the items in non-decreasing order by size, as we do for *FFD*. We then pick some obvious lower bound q on $OPT(L)$ and imagine we have q empty bins, $B_1, B_2, ..., B_q$. Place p_1 in B_1 and proceed through the list of items, packing p_i in a bin whose current contents has minimum total size (breaking ties by bin index, when necessary). If we ever reach a point where p_i does not fit in *any* of the q bins (either because the capacity C or the limit m is exceeded), we halt the iteration, increase q by 1, and start over. Eventually we will succeed in generating a packing for some value of q, and this will be the output.

The running time of *ILFD* is $O(n^2 \log n)$, but this can be improved to $O(n \log^2 n)$ by using binary search on q. The performance bound proved for *ILFD* is $R_{ILFD}^{\infty} \leq 2$, which makes *ILFD* competitive with *FFD*. It is conjectured that the actual value of R_{ILFD}^{∞} is closer to the 4/3 value we cited in the last section for the case when there is no limit on the number of items per bin.

Partial Orders on L — Partial orders, $\leq$, on the set L of items arise in two potential applications of bin-packing. One is again related to multiprocessor scheduling, and was studied by Garey, Graham, Johnson and Yao in [50]. Suppose we have a set of unit-length tasks $p_1, ..., p_n$ with resource requirements subject to an over-all bound of C, as above, but with no limit on the number of processors. In this case a partial order $\leq$ is interpreted as follows: $p_i \leq p_j$ means that p_i must be executed *before* p_j, i.e., must be assigned to a bin with lower index than that to which p_j is assigned.

The other application in which partial orders arise is in "assembly line balancing," and is studied by Wee and Magazine [107]. Here the items represent tasks to be performed on a single product as it moves along an assembly line. Each is performed at one of a sequence of workstations B_1, B_2, etc., and the item sizes correspond to the times required to execute the tasks. The assembly line advances in discrete steps, stopping for a period of time C at each workstation. Thus a set of tasks can be assigned to a work station (bin) if their total time (size) does not exceed C. The goal is to minimize the number of workstations required. In this case a partial order $\leq$ has the following interpretation: $p_i \leq p_j$ means that in any assignment of tasks to workstations (bins), p_i must be performed before p_j (but they could be performed at the same workstation, merely by doing p_i before p_j within the total time C allowed, so this time p_i can go either in an earlier bin or the *same* bin as p_j).

Note that these two applications yield different interpretations of the partial order constraint within the bin packing context. See Figure 4. Although this difference might appear to be slight, its consequences, as shown in the figure, are nontrivial. The algorithm referred to there, ORDERED FIRST FIT DECREASING, is the best algorithm known for either version of the problem, but yields quite different guarantees. It is quite simple to describe. First, we order the items by non-increasing size, as with FFD. We then pack bins, rather than items, in sequence. Bin B_i is packed as follows: Place the largest unpacked item into B_i that the partial order will allow. Repeat until no more items can legally be packed into B_i.

Note that this algorithm can be applied to either version of the problem, so long as the partial

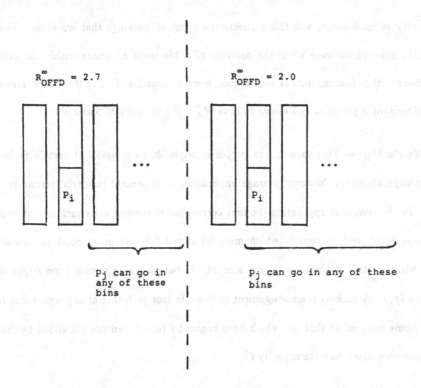

$$R^\infty_{OFFD} = 2.7 \qquad\qquad R^\infty_{OFFD} = 2.0$$

P_i

P_i

P_j can go in any of these bins

P_j can go in any of these bins

Figure 4. Two interpretations of $P_i \lesssim P_j$ and their consequences.

order is interpreted appropriately. Note also that, in the absence of a partial order, this algorithm generates the same packing as FIRST FIT DECREASING and hence has an asymptotic worst-case ratio of 11/9.

Clustered Items — Here, the basic idea is that only "closely related" items may go in a bin together. The one example we cite is from a paper by Chandra, Hirschberg, and Wong [18], although other potential applications of this type might come to mind. Here the items are thought of as having geographical locations. Putting them in the same bin corresponds to assigning them to a common facility (computing service, telephone switching center, etc.), where each such facility is assumed to have a standard capacity C. We desire that the items which are served by a common facility be in close proximity to each other, and this restricts the types of packings that we allow. The main results in [18] concern the case when the contents of a bin must all reside within the same unit square, although other figures, such as unit circles, are also considered. For the unit square case, a geometric algorithm is proposed and shown to have R_A^∞ lying between 3.75 and 3.8.

Dynamic Bin-Packing — This variant was proposed originally as a model of certain problems in computer storage allocation; however, storage applications in a general industrial setting are easily envisioned. In the computer application the bins correspond to storage units such as disk cylinders, and the items correspond to records which must be stored for certain specified periods of time. Associated with an item is thus not only a size $s(p_i)$, but also a beginning time $b(p_i)$ and an ending time $e(p_i)$. A packing is an assignment of items to bins such that at any time t and for any bin B, the items assigned to that bin which have begun by time t and not yet ended by that time have total size no greater than the capacity C.

The research reported by ourselves in [27] concentrates on "on-line" algorithms, where in this case an on-line algorithm packs items in the order in which they begin, and may not use information about items which are to begin later, or the ending times for items which are currently in the packing (this lack of information mirrors the predicament often faced by actual computer storage allocators). It is assumed that once an item is assigned to a bin it cannot be moved to another bin.

The algorithm FIRST FIT can be readily adapted to this situation, but the dynamic nature of the environment significantly impairs its performance. For the case when no item size exceeds $C/2$, we have $R_{FF}^\infty(1/2) = 1.5$ in the classical case, but in the dynamic case it is shown in [27] that *any* on-line algorithm must obey $R_A^\infty(1/2) \geqslant 5/3 = 1.666...$. For FIRST FIT it was proved that $R_{FF}^\infty(1/2)$ lies somewhere between 1.75 and 1.78. The case when items larger than $C/2$ are allowed is even more difficult to analyze (as seems usually to be the case with bin packing), but is clearly much worse. Here it is known that R_{FF}^∞ lies somewhere between 2.75 and 2.89, and *any* on-line algorithm must obey $R_A^\infty \geqslant 2.5$.

Studies of dynamic packings are at the interface between bin-packing and dynamic storage allocation. The latter class of problems is distinguished by the assumption that items, once packed, can not be moved at all prior to their departure. (In dynamic bin packing the allowed movement of items *within* bins was implicit; only the movement of an item from one bin to another was disallowed.) Under this added assumption the fragmentation (alternating holes and occupied regions) that develops as items come and go can create far more wasted space; the space wasted by the partitioning of storage into "bins" is usually minor by comparison. The standard model considers only a single bin whose capacity is to be determined under a given algorithm and the assumptions: the total size of items present at any time never exceeds m and the maximum item size is j. (Note that m is our usual lower bound on the capacity required by an optimal algorithm.)

FIRST FIT and BEST FIT are the principal approximation algorithms that have been studied. With FIRST FIT an arriving item is stored at the beginning of the first sufficiently large hole encountered in a scan of the bin. BEST FIT is defined similarly, where the hole is a smallest one exceeding the item size. Let $C_A(j,m)$ denote the capacity needed in the worst-case under algorithm A, and in the spirit of our other asymptotic bounds let $C_A^\infty(j) = \lim_{m \to \infty} C_A(j,m)/m$.

Robson [99] has shown that $\frac{1}{2} \log_2 j \leqslant C_{FF}^\infty(j) \leqslant \log_2 j$ and $C_{BF}^\infty(j) = \Theta(j)$. The bound for FIRST FIT is a best possible one in the sense that an optimal algorithm must have a $\Theta(\log j)$ worst case if, like FF and BF, it must allocate storage to each item at its time of arrival, and if no information is available on items that have not yet arrived. This last result of Robson [98] is in fact the classical one of dynamic storage allocation, and generalizes earlier results of Graham [61].

There are many other interesting results for this problem, particularly those specializing item sizes to powers of two. Knuth [83] covers the elements of the subject and a recent survey appears in [23].

4. Multiprocessor Scheduling

We continue here with the bin-design problem defined in the Introduction: With the number of bins fixed, find the smallest, common bin capacity C sufficient to pack L.

The initial work on approximation algorithms for this problem appears in [60,62]. We can define worst case ratios as we did before, noting that in this case the asymptotic and absolute worst case ratios will coincide for reasonable algorithms — any worst case example can be converted to one with an arbitrarily large value for C merely by scaling up all the sizes by an appropriate multiplicative factor. (In those applications in which there is a fixed upper bound on the possible item sizes, asymptotic worst case bounds would make sense, but due to the nature of the problem we would tend to get $R_A^\infty = 1$ for most algorithms, e.g., see [80]). Thus we shall express results for this problem in terms of the absolute ratio R_A. Graham examined two basic algorithms. LOWEST FIT assigns the items to bins in order, placing p_i in a bin with current contents of minimum total size (ties broken by bin index when necessary). LOWEST FIT DECREASING first sorts the items so that they are in non-increasing order by size and then applies LOWEST FIT. Fixing m, the number of bins, Graham was able to prove that $R_{LF} = 2-(1/m)$ [60] and that $R_{LFD} = (4/3)-(1/3m)$ [62]. In analogy with the results for $R_A^\infty(t)$ in the classical problem, Coffman and Sethi [35] showed that the LFD bound improves to $R_{LFD}(k) \leq \dfrac{k+1}{k} - \dfrac{1}{km}$ when it is known that there are at least $k \geq 3$ items per bin. In [88] Langston demonstrated how the bad examples of LFD could be effectively avoided. The resulting algorithm, LFD^*, uses an iterative technique and has a worst-case bound, $R_{LFD^*} \leq \dfrac{5}{4} + \dfrac{1}{12}(2^{-k})$, where k is the number of iterations chosen.

Sahni [100] developed an approximation scheme for this problem several years ago. In particular, he showed that for any fixed value of m and any $\epsilon > 0$ there exists a polynomial time algorithm with $R_A = 1+\epsilon$. As in the corresponding studies for the bin-packing problem discussed earlier, the result is primarily of theoretical interest; such algorithms are exponential in m (and polynomial in $1/\epsilon$) and therefore unattractive for $m > 3$ or very small ϵ.

A more practical algorithm which is polynomial in m and still improves upon LFD was presented by ourselves in [26]. The algorithm works on an iterative principle much like that of $ILFD$. Called MULTIFIT DECREASING, the algorithm works by guessing a capacity C and then applying FFD to the list. The next guess is either larger or smaller, depending on whether FFD used more than m bins of that capacity to pack the list or not. By using an appropriate binary search strategy and limiting the number of iterations performed to some small number k, we obtain an algorithm, denoted $MF(k)$, for which we proved the bound $R_{MF[k]} \leq 1.220+(1/2)^k$. Friesen [44] subsequently established the improved upper bound, $6/5 + (1/2)^k$, which again is independent of m. $MF(k)$ improves upon LFD for all $m > 2$ when $k \geq 5$, at the cost of only a small increase in running time. The worst behavior known for this algorithm is shown in examples constructed by Friesen [44], which imply that $R_{MF[k]} \geq 13/11 = 1.18181...$ for $m \geq 13$. (Better upper bounds are known for the cases when $m \leq 7$ [26]).

We remark in passing that the results for MULTIFIT DECREASING are proved by considering the following bin packing variant: Suppose we are given two sets of bins, one with bins all of capacity α, the other with bins all of capacity β. What is the asymptotic worst case ratio of the number of β-capacity bins used by FF (or FFD) to the minimum number of α-capacity bins needed? This question, first raised in [49], is investigated in detail in Friesen's thesis [44] as well as in [26], and is used in [49] for proposing a conjectural explanation of the mysterious fraction 17/10 in the original theorem about R_{FF}^∞ (a conjecture that, unfortunately, is only partially true [103]).

An algorithm structurally different from any considered so far was introduced by Finn and Horowitz [41] and later improved by Langston [89]. It is based on finding approximate solutions to the following closely related partitioning problem: Partition L into m blocks (bins) $B_1,...,B_m$ so as to minimize the difference in the maximum and minimum bin levels, viz.

$$D(B_1,...,B_m) = \max_i\{\ell(B_i)\} - \min_i\{\ell(B_i)\} ,$$

where $\ell(B_i) = \sum_{p \in B_i} s(p)$. As is to be expected, good approximate solutions for this problem are good approximate solutions for the problem of minimizing bin capacity. (The two problems are equivalent, of course, for $m = 2$.) The basic idea of the algorithm is iteratively to exchange items in the two bins having the maximum and minimum levels until D can no longer be reduced in this way.

As shown in [41] linear time algorithms can produce packings comparing favorably with those of MULTIFIT DECREASING, especially for large lists. Quite recently, however, Karmarkar and Karp [75] obtained even stronger results. Again concentrating on better performance for a large number, n, of items, they devised algorithms based on an operation called set differencing. These algorithms are best illustrated for the case $m = 2$.

The differencing operation consists of selecting a pair of item sizes $s(p)$ and $s(p')$ from L and then restricting the solution to partitions in which p and p' appear in different bins. The new, smaller problem is then equivalent to partitioning

$$L' = L - \{s(p), s(p')\} \cup \{|s(p) - s(p')|\} .$$

Consider for example, the algorithm: *While $|L| > 1$, iteratively select the largest two elements s and s' of L and perform the operation $L \leftarrow L - \{s, s'\} \cup \{|s-s'|\}$.* The last number, when $|L| = 1$, determines $D(B_1, B_2)$; the corresponding partition is trivial to construct by backtracking through the sequence of differencing operations. Figure 5 shows an example in the form of a tree. Other set-differencing algorithms can be obtained simply by altering the order in which elements are selected for differencing.

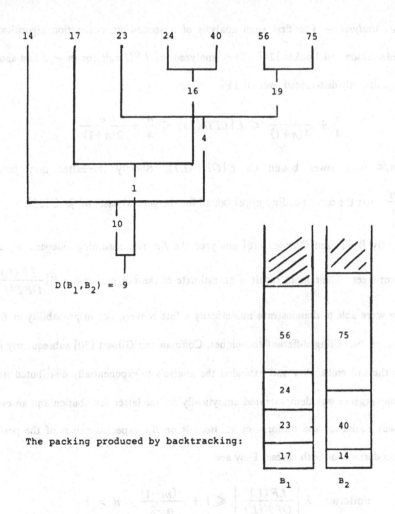

Figure 5.   An Illustration of Set-Differencing.

The methods for $m = 2$ are extended in a natural way to classes of algorithms for arbitrary $m > 2$. For a particular linear-time algorithm Karmarkar and Karp show that, except in pathological cases, partitions are produced with $D = O(n^{-\log n})$. This result is to be contrasted with the corresponding $O(n^{-1})$ result that applies to LFD and MULTIFIT. To show that the pathological cases are extremely rare they use a simple probability model to verify that as $n \to \infty$ the algorithm performs as claimed with probability 1. We shall return briefly to this analysis later on.

Average-Case Analysis — The first such analysis of a specific approximation algorithm was by Coffman, Frederickson and Lueker [24]. They analyzed the *LFD* rule for $m = 2$ and showed that for item sizes uniformly distributed over [0,1]

$$\frac{n}{4} + \frac{1}{4(n+1)} \leqslant E[LFD(L)] \leqslant \frac{n}{4} + \frac{e}{2(n+1)} .$$

Note that $n/4$ is a lower bound on $E[OPT(L)]$. Shortly thereafter they proved that

$\frac{n}{2m} + O(\frac{m}{n})$ was the corresponding upper bound for the general case, $m \geqslant 2$ [25].

More recently Bruno and Downey [16] analyzed the *LF* rule assuming independent, uniformly distributed item sizes. Their main result is an estimate of the tail probability $P\{\frac{LF(L)}{OPT(L)} > x\}$ by which they were able to demonstrate numerically a fast convergence in probability of $LF(L)$ to $OPT(L)$ as $n \to \infty$. Using different techniques, Coffman and Gilbert [30] subsequently improved the bound on the tail probability and extended the analysis to exponentially distributed item sizes. Exponential convergence was demonstrated analytically for the latter distribution and an even faster convergence was found for the uniform case. Bounds on the expected values of the performance ratios were also derived for both cases. They are

$$\text{uniform:} \quad E\left[\frac{LF(L)}{OPT(L)}\right] \leqslant 1 + \frac{2(m-1)}{n-2}; \quad n > 2$$

$$\text{exponential:} \quad E\left[\frac{LF(L)}{OPT(L)}\right] \leqslant 1 + \frac{(m-1)H_{m-1}}{n-m}, \quad n > m ,$$

where H_i is the ith harmonic number.

Asymptotic results of a more general nature were proved by Dempster et al. [36]. They showed that for any approximation algorithm, A, in a broad class (including all those considered in this survey), the ratio of $A(L)$ to $OPT(L)$ convergences in probability to 1 as $n \to \infty$, if the item sizes are independent and identically distributed random variables with finite variance. Loulou [92] subsequently proved a stronger result for the LF and LFD rules, viz. the absolute errors $LF(L)-OPT(L)$ and $LFD(L)-OPT(L)$ converge in probability to finite random variables.

Quite recently Frenk and Rinnooy Kan [43] dealt successfully with a conjecture of Loulou; they proved that $LFD(L)-OPT(L)$ converges in probability to 0 as $n \to \infty$ if the item size distribution has a finite mean and a density f satisfying $f(0) > 0$. They also show that if the distribution is uniform or exponential, the rate of convergence is $O\left[\dfrac{\log n}{n}\right]$.

What appear to be the strongest results of this type currently known were obtained recently by Karmarkar and Karp [75] for one of the set-differencing algorithms discussed earlier in this section, where the objective function is again the difference D between the largest and smallest bin levels. For the particular algorithm used in [75] let $D^*(L)$ denote the output for a given m, and define

$$D^*_{m,n} = \sup_{\{L: |L| \leqslant n\}} D^*(L) .$$

They showed that there is a positive constant α such that

$$D^*_{m,n} = e^{-\alpha(\ell n\, n)^2/m}$$

with probability 1 as $n \to \infty$ for any item size distribution satisfying a mild smoothness condition.

A similar result is conjectured for the simple largest-pair-first algorithm illustrated in Fig. 5. However, a rather more elaborate, but still linear-time algorithm had to be adopted in order to avoid the usual difficulties in dealing with order statistics, as well as certain other problems.

To help motivate the above result let us assume $m = 2$, where the simpler form $D^*_{2,n} \leqslant O(n^{-\alpha \log n})$ applies. Consider the following set-differencing algorithm which assumes that $n = 2^k$, $k \geqslant 1$. Pair the largest two numbers in L, the next largest two, and so on. Differencing each pair we establish a reduced problem containing $n/2 = 2^{k-1}$ differences. The desired partition is obtained from this process repeated k times. Under general assumptions it is reasonable to expect that at the end of each of the k stages the order of magnitude of the numbers is reduced approximately by a factor of n. Accordingly, we expect a final partition such that $D(L) = O(n^{-k})$ or $D(L) = O(n^{-\log n})$.

A result in a similar vein was recently proved by Karmarkar, Karp, Lueker and Odlyzko [77] for a problem posed by Michael Steele [105]. The problem was to find the rate of convergence of $E[D(L)]$ as $n \to \infty$ assuming an optimal algorithm, $m = 2$ and item sizes uniformly distributed over $[0,1]$. Using a technique called the second-moment method, they showed that $P[D(L) > n^2 2^{-n}] = O(n^{-1})$.

5. Other Performance Criteria

Sums of Squares of Bin Levels — In parallel efforts Cody and Coffman [21] and Chandra and Wong [17] studied bin packing problems arising in the allocation of records on computer auxiliary storage devices. The basic probability models are patterned after those analyzed by Knuth [83] in connection with similar storage assignment problems. In bin packing terms the "sizes" of items are the access frequencies of the corresponding records.

The problem in [21] models paging drums where a given set of pages is to be partitioned among the m sectors (bins) of the drum so as to minimize average access time. This quantity is calculated to be $\dfrac{m-2}{2} + \dfrac{m}{2} \sum_{i=1}^{m} \ell^2(B_i)$, where $\ell(B_i)$ is the ith bin level, i.e. the sum of the access frequencies of the pages in the ith sector. The minimization of this sum can be accomplished roughly by making the bin-levels all as close to each other as possible. In [21] the *LFD* rule is applied to this problem and the following result proved: $R_{LFD} \leqslant 1 + \dfrac{1}{16(m-1)}$.

The problem in [17] models arm contention in disk-pack computer storage. In this case, the object is to minimize the contention that occurs whenever two items from the same bin are requested at the same time. Contention is measured here by the simpler quantity, $\sum\limits_{i=1}^{m} \ell^2(B_i)$. The algorithm *LFD* is analyzed in this context too, and it is shown that for this problem $\frac{37}{36} \leqslant R_{LFD} \leqslant \frac{25}{24}$. In [38], Easton and Wong consider the variant in which no bin can contain more than k items. They analyze an appropriately modified version of *LFD* and showed that $R_{LFD} \leqslant 4/3$.

Wong and Yao [108] consider yet another variant based on minimizing access time [110]. Suppose we wish to *maximize*, rather than minimize, the sum of squares in the above case where k items per bin are allowed. This might be considered as a bin packing problem where all items have both a size $s(p) = 1$ and an arbitrary weight $w(p)$, the bin capacity is k, and the goal is to pack the items into m bins so as to maximize the sum of the squares of the total weight in each bin. As observed in [110], this is of course an easy matter: merely put the k largest items in the first bin, the next k largest in the second bin, etc. Wong and Yao consider the generalization where the sizes as well as the weights are arbitrary.

In order that results for this maximization problem can be compared directly to those for the minimization problems we have been studying so far, we shall define $R_A(L)$ to be $OPT(L)/A(L)$ for any approximation algorithm A (this is the inverse of our definition for minimization problems). R_A and R_A^∞ are then defined as before and lie in the range $[1,\infty)$. Wong and Yao propose a heuristic based on ordering the items by non-decreasing density (weight divided by size) and then applying NEXT FIT. They show that this heuristic satisfies $R_A \leqslant 2$.

Maximizing the Number of Items Packed — We consider again a maximization problem that fixes the number of bins and the bin capacity. This time the goal is to pack as many of the items in L as possible into the bins. Coffman, Leung, and Ting [34] consider the algorithm FIRST FIT INCREASING, which first sorts the items into non-decreasing order by size, and then applies *FF* until an item is reached which will not fit in any of the bins (which implies that none of the remaining items will fit either). They show that $R_{FFI} = 4/3$. In [33], Coffman and Leung consider an algorithm that, like *ILFD* and *MFD*, involves iteration. Their algorithm, denoted *FFD**, works as follows: First sort the items in non-increasing order by size, and then apply *FF*. If some item fails to fit, stop, delete the first (largest) item in the list, and reapply *FF* to the shorter list. Repeat this until a list is obtained that *FF does* pack into the *m* bins. Coffman and Leung show that *FFD** will always pack at least as many items as *FFI*, and indeed obeys the better bounds $8/7 \leqslant R_{FFD^*}^\infty \leqslant 7/6$, making the added complexity of *FFD** over *FFI* worth the effort. Langston [87] has recently analyzed these heuristics for the more general model in which bin sizes may vary. By arranging bin sizes in non-decreasing order, he proves that $R_{FFI}^\infty = 2$ and $11/8 \leqslant R_{FFD^*}^\infty \leqslant 3/2$.

Maximizing the Number of Bins above a Given Level — Suppose a threshold $T > 0$ is given. Assmann, Johnson, Kleitman and Leung [3] studied the problem of finding a packing of *L* into a maximum number of bins such that each bin has a level not less than *T*. Similar to the sum of squares problem, for a good packing, the goal is roughly to pack every bin to a level as close as possible to, but not less than *T*.

Two approximation algorithms were examined for this problem. The first that we shall describe begins by producing a standard *FFD* packing of *L* for some given capacity $C > T$. The second stage iteratively takes items from the last non-empty bin and places them in the currently lowest indexed bin having a level less than *T*; at the end of this stage some, possibly empty subset of highest indexed bins in the *FFD* packing will have been emptied in order to bring the levels of lower indexed bins up to at least *T*. The performance guarantee for this rule, called *FFD[C]*, is an interesting function of the value of *C* chosen for the initial *FFD* stage. Using our inverted ratio $OPT(L)/A(L)$ as before, it is shown in [3] that $R_{FFD(C)}^\infty \geqslant 3/2$ for all $C \geqslant T$ and

$\lim_{C \to T} R^{\infty}_{FFD[C]} = 2$ and $\lim_{C \to 2T} R^{\infty}_{FFD[C]} = 2$, whereas $R^{\infty}_{FFD[C]} = 3/2$ if $\frac{4}{3}T \leqslant C < \frac{3}{2}T$. In other words, for best worst-case performance we should choose C in $[\frac{4}{3}T, \frac{3}{2}T)$.

The second algorithm investigated was simply ITERATED LOWEST FIT DECREASING adapted to this problem. A value is guessed for the number, m, of bins in which to apply LFD. If each bin in the LFD packing has a level at least T, the algorithm halts. Otherwise, a smaller value of m is taken and the procedure repeated. It is easily verified that an efficient binary search can be organized around the fact that an appropriate m must exist in the range $\left[\frac{1}{2T} \sum_{i=1}^{n} s(p_i), \frac{1}{T} \sum_{i=1}^{n} s(p_i) \right]$. The corresponding algorithm has the asymptotic bound $R^{\infty}_{ILFD} = 4/3$, thus improving on the bound for $FFD[C]$.

An experimental analysis of average case behavior for these algorithms, plus a probabilistic analysis of NF analogous to that in [72] can be found in [2].

Maximizing the minimum bin level — This dual to the capacity minimization problem was studied by Deuermeyer, Friesen and Langston [37]. Clearly, it is also closely related to the problem of minimizing the difference between maximum and minimum bin levels, which was discussed in the previous section. In [37] it is shown that the LFD rule has a 4/3 bound (using the inverted ratio) for the max-min problem, just as it has for the min-max problem. As might be expected, set-differencing algorithms are also effective for this problem, but algorithms such as MULTIFIT can perform very poorly.

6. Vector Packing

In this section we consider one way of generalizing the classical one-dimensional bin packing problem to higher dimensions. Instead of each $s(p)$ being a single number, we consider the case when it is a d-dimensional vector $s(p) = <s_1(p), s_2(p), ..., s_d(p)>$. The bin capacity is also a

d-dimensional vector $<C,C,...,C>$, and the goal is to pack the items in a minimum number of bins, given that the contents of any given bin must have vector sum less than or equal to its capacity. This problem models multiprocessor scheduling of unit-length tasks in the case when there are d resources, rather than just one as we assumed before. For simplicity we have normalized the amounts of resources available so that all d bounds are the same.

Note that the two-dimensional version of this problem is not the same as the problem of packing rectangles (to be discussed in the next section). A vector $<s_1(p),s_2(p)>$ could be thought of as representing a rectangle with length $s_1(p)$ and width $s_2(p)$, and a bin of capacity $<C,C>$ as a square into which the rectangles are to be packed. However, the only types of packings allowed here would correspond to ones in which the rectangles were placed corner to corner, diagonally across the bin.

In [84], Kou and Markowsky show that any "reasonable" algorithm, i.e., one which does not yield packings in which the contents of two non-empty bins can be combined into a single bin, obeys the bound $R_A \leqslant d+1$, where d is the number of dimensions (an alternative proof can be found in [48], although there the theorem is not stated in its full generality). We note in passing that, in spite of the obvious desirability of the above "reasonable" property, not all the algorithms we have mentioned so far are reasonable — an obvious offender is NEXT FIT. However, FIRST FIT, FIRST FIT DECREASING, and many others *are* reasonable and hence do obey the above-mentioned, not very impressive (when d is large), bound. They are, in fact, better than reasonable, but not by as much as one would like. In [50], Garey, Graham, Johnson, and Yao analyze the d-dimensional problem and appropriate adaptations of FF and FFD to this multi-dimensional case (in FFD the items are sorted in non-decreasing order by the maximum components of their size vectors). They show that $R_{FF}^{\infty} = d+7/10$, which reduces to the familiar 17/10 result in the one-dimensional case, and that $d \leqslant R_{FFD}^{\infty} \leqslant d+1/3$. To date, no one has found any polynomial time approximation algorithm for the general d-dimensional problem with $R_A^{\infty} < d$. Yao has shown

[108] that any algorithm that is *faster* than *FF* or *FFD*, i.e., that has a running time that is $o(n \log n)$ in the decision tree model of computation, *must* have $R_A^\infty \geqslant d$. Within this constraint, Fernandez de la Vega and Leuker [40] show, by extending their results in one dimension, that a polynomial time approximation scheme exists for the vector problem as well; i.e. there is a linear-time algorithm for finding solutions within ϵ of d times the optimal. (The earlier caveats concerning running time apply here as well, however.) In spite of these results, we should note that the extensive simulation results of Maruyama, Chang, and Tang [96] for *FF*, *FFD*, and a variety of other algorithms indicate that average case behavior may not be nearly so bad here as the worst case bounds.

In the variant on this problem in which a partial order is present, however, things definitely get worse. Suppose that the set of items has a partial order $\leq$ associated with it that constrains the allowable packings as in Section 3 (the multiprocessor scheduling rather than the assembly line balancing case). In this case the natural generalization of the ORDERED FIRST FIT DECREASING algorithm of Section 3 can be shown [50] to obey $(1.691)d+1 \leqslant R_{OFFD}^\infty \leqslant (1.7)d+1$, a definite worsening of our bounds when no partial order was present (the result mentioned in Section 3 for the one-dimensional version of this problem is a special case of this result). Similar results are obtained for the algorithm ORDERED FIRST FIT BY LEVEL, which works the same way as OFFD, except that instead of ordering items by non-increasing maximum size component, they are ordered by non-increasing "level" in the partial order [24]: $R_{OFFL}^\infty = (1.7)d+1$. That some type of pre-ordering is necessary for even this standard of performance follows from the fact that the algorithm without any pre-ordering, ORDERED FIRST FIT, has $R_{OFF}^\infty = \infty$ [50].

7. Rectangle Packing

In this section we consider an active area of bin packing research: the problem of packing rectangles into two-dimensional bins. The first version of this problem to be studied from a performance guarantee point of view is due to Baker, Coffman, and Rivest [8], and models a variety of problems, from computer scheduling to stock cutting. In this version, the items p_i are rectangles, with height h_i and width w_i. The goal is to pack them in a vertical strip of width C, so as to minimize the total height of the strip needed. The rectangles must be packed orthogonally, that is, no rotations are allowed: all rectangles must have their width parallel to the bottom of the strip.

The orthogonality restriction is justified on the basis of the proposed application to scheduling. Here the items once again correspond to tasks. The height of an item is the amount of processing time it requires, and its width is the amount of contiguous memory it needs. The strip width C is then the total memory available; the strip length is the amount of time needed to schedule all the items. In this application it makes no sense to rotate a rectangle, even by ninety degrees, since execution time is not in general directly translatable into a memory requirement.

Applications to stock-cutting occur in a variety of industrial settings where the "raw" material involved comes in rolls, for instance rolls of paper, rolls of cloth, rolls of sheet metal, etc. From these rolls we may wish to cut patterns (for labels, clothes, boxes, etc.) or merely just shorter, narrower rolls. In the simplest case, we can view the objects we wish to cut from the rolls as being, or approximating, rectangles. We minimize our wastage if we minimize the amount of roll (the strip length) used. Once again some form of orthogonality may be justified, since in many applications, the cutting is done by blades that must be either parallel or perpendicular to the strip, and the material may have a bias that dictates the orientation of the rectangles. However, ninety degree rotations may in some cases be allowable, and we will later say a bit about how the results we discuss can be extended to take this into account.

Because of the economic importance of efficient stock-cutting, a broad range of classical heuristic and enumerative methods have been applied in the last 20 years. For example, solution techniques have been designed around linear programming, dynamic programming, branch and bound, network flow and heuristic search methods. (See [1,10,19,57] for such studies and references to a number of others.) The performance of these solution techniques is normally evaluated experimentally, rather than analytically, so they fall outside the scope of this survey.

In [8], Baker, Coffman, and Rivest consider a variety of strip packing algorithms based on a "bottom up — left justified" (BOTTOM-LEFT for short) packing rule. In a BOTTOM-LEFT packing, items are packed in turn, each item being placed as near to the bottom of the strip as it will fit and then as far to the left as it can be placed at that bottom-most level. Note that there is a difference in kind between two-dimensional packing rules, such as the BOTTOM-LEFT rule, and one-dimensional rules such as FIRST FIT and NEXT FIT. In the one-dimensional case there always exists an ordering of the items such that FIRST FIT (NEXT FIT) constructs an optimal packing. However, as shown in [8] this is not the case for BOTTOM-LEFT. In fact, Brown [13] has constructed instances in which the best BOTTOM-LEFT packing possible still yields a strip whose height is 5/4 times optimal.

However, although no preordering of the items may be able to yield an optimal packing, some may still be better than others. Various BOTTOM-LEFT algorithms can be considered, depending on how (if at all) the set of rectangles is initially preordered. It turns out that only one of the standard orderings seems to make a difference as far as worst case behavior is concerned. If we let BL stand for the simple BOTTOM-LEFT algorithm, and $BLIW$, $BLIH$, $BLDW$, and $BLDH$ stand for the algorithm with preordering by increasing (i.e., non-decreasing) width, increasing height, decreasing width, and decreasing height, then we have

$$R_{BL} = R_{BLIW} = R_{BLIH} = R_{BLDH} = \infty; \quad R_{BLDW} = 3$$

For the special case of squares ($h_i = w_i$ for all p_i), the $BLDH$ algorithm becomes equivalent to $BLDW$, and the result improves to $R_{BLDW} = 2$ [8].

For the case of arbitrary rectangles, subsequent work has yielded some improvements. The FIRST FIT DECREASING HEIGHT "level" algorithm of Coffman, Garey, Johnson, and Tarjan [29] (to be described later) can be shown to have $R_{FFDH} = 2.7$, and an algorithm of Sleator [104] further reduced the bound to $R_A = 2.5$.

These results all concern *absolute* worst case performance ratios. Indeed, for this problem it would again seem as if absolute and asymptotic performance ratios should be equivalent, since heights can be scaled to arbitrarily large values. However, such scaling may not be sensible in many practical applications, where some strict upper bound on height may be imposed. In this case, asymptotic analysis may be a more meaningful measure, giving us guarantees that hold as the optimal strip length becomes very large with respect to this maximum possible item height. As might be expected, these asymptotic guarantees can be better than the absolute ones (although they do not equal 1, as they would if all rectangle widths were equal, thus reducing us to the capacity minimization problem of Section 4). For instance, $R_{BLDW}^{\infty} = 2$, an improvement of 1 over the absolute guarantee for *BLDW*, but a long way from optimal.

The search for strip packing algorithms with better asymptotic worst case ratios was taken up in [29] by Tarjan and ourselves. The new algorithms were based on a different type of packing rule, suggested by Golan [58], and were called "level" algorithms. These algorithms involve an attempt to apply our old knowledge about one dimensional bin packing. Note that if all rectangles have the same height, the two-dimensional problem essentially reduces to the one-dimensional case: in an optimal packing the items may be placed in rows or "levels." Each level in the packing then corresponds to a bin and the height of the packing corresponds to the number of bins used. The basic idea of a level algorithm is the following: First, the items are preordered by non-increasing height. The packing is then constructed as a sequence of *levels*, each rectangle being placed so that its bottom rests on one of these levels. The first level is simply the bottom of the bin. Each subsequent level is defined by a horizontal line drawn through the top of the tallest rectangle on the previous level. This is best illustrated by considering the two basic level algorithms proposed in [29].

In the algorithm NEXT FIT DECREASING HEIGHT rectangles are packed left-justified on a level until the next rectangle will not fit, in which case it is used to start a new level above the previous one, on which the packing proceeds. Note the analogy with the one-dimensional NEXT FIT algorithm. In the FIRST FIT DECREASING HEIGHT algorithm (another analog), each rectangle is placed left-justified on the first (i.e., lowest) level in which it will fit. If none of the current levels has room, a new one is started as with the *NFDH* algorithm. See Figure 6 for an example of an *FFDH* packing.

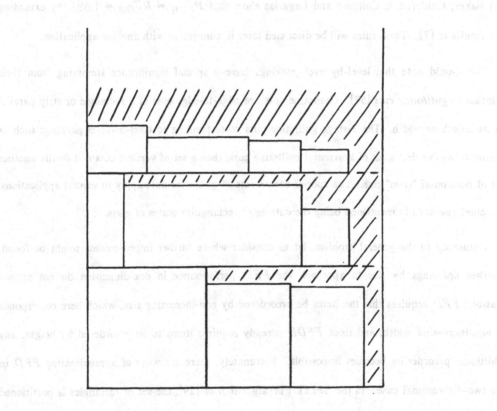

Figure 6. An example of FFDH packing.

At first glance, one would expect *NFDH* and *FFDH* to be worse than their one-dimensional counterparts, given all the space that may be wasted in a level above items which are shorter than the first one. However, it turns out that this wasted space is strictly bounded, and by a collapsing sum argument it can be concluded that, exactly as in the one-dimensional case, $R_{NFDH}^{\infty} = 2$ and $R_{FFDH}^{\infty} = 1.7$. The results for bounded item widths also resemble their one-dimensional counterparts.

For the special case of squares the asymptotic worst case ratio is reduced to 1.5. The INCREASING rules BLIH (or BLIW) and NFIH have also been analyzed for this special case. In [6] Baker, Calderbank, Coffman and Lagarias show that $R_{BLIH}^{\infty} = R_{NFIH}^{\infty} = 1.691$, by extending the results in [7]. These rules will be discussed later in connection with another application.

We should note that level-by-level packings have a special significance stemming from their relation to *guillotine cuts* [57]. Guillotine cuts are edge-to-edge cuts of a rectangle or strip parallel to its length or width. The 3-stage guillotine cuts corresponding to level-by-level packings such as Figure 6 involve first a set of horizontal guillotine cuts, then a set of vertical cuts and finally another set of horizontal "trim" cuts. The special constraints of guillotine cuts apply in several applications, the chief one usually mentioned being the cutting of rectangular plates of glass.

Returning to the general problem, let us consider where further improvements might be found. Further orderings by size to approach the *FFD* performance in one-dimension do not appear feasible; *FFD* requires that the items be preordered by non-increasing size, which here corresponds to non-increasing width, and since *FFDH* already requires items to be preordered by height, any additional preordering becomes impossible. Fortunately, there are ways of approximating *FFD* in the two-dimensional case. In the SPLIT FIT algorithm of [29], the set of rectangles is partitioned into two parts, those with width exceeding $C/2$ and those without, and each subset is ordered by non-increasing height. Packings for the two sets are then combined in an involved manner, and the result is an algorithm with $R_{SF}^{\infty} = 1.5$. This idea of splitting the set of rectangles into subsets according to width can be carried even further. In [59] Golan described an algorithm for which $R_A^{\infty} \leqslant 4/3$, and in [5] Baker, Brown, and Katseff devised a much more complicated algorithm for which $R_A^{\infty} \leqslant 5/4$, a bound which is close to the 11/9 guarantee provided by *FFD* in the one-dimensional case.

So far, all the rectangle packing algorithms we have discussed for which $R_A^{\infty} \leqslant \infty$ have involved some preordering of the rectangles, and hence are not "on-line" algorithms. However, such algorithms might well be required in scheduling applications, and so the question of finding an on-line algorithm with reasonable worst case behavior becomes relevant. Baker and Schwartz, in [9], show that such algorithms exist by devising what they call "shelf" algorithms. These are variants on the level algorithms above in which levels, rather than being determined by their tallest item, come in fixed sizes. If we assume that 1 is an *a priori* upper bound on rectangle height, the standard levels will come in heights r^{-k}, $k \geqslant 0$, for some prespecified value of r, $0 < r < 1$. Whenever a rectangle p_i is to be packed in the NEXT FIT SHELF(r) algorithm, one first determines that value of k such that $r^{k+1} < h_i \leqslant r^k$. If there is a level of height r^k already in the packing, and p_i will fit in the currently active one, it is placed there. Otherwise it is placed in a new such level, which becomes the currently active one for that height. The algorithm FIRST FIT SHELF(r) is defined analogously.

Although these shelf algorithms clearly have considerable space-wasting potential, it turns out that the wastage can once again be bounded, and in fact $R_{NFS(r)}^{\infty} = 2/r$ and $R_{FFS(r)}^{\infty} = 1.7/r$. Note that these approach the values for *NFDH* and *FFDH* as r approaches 1. However, as r approaches 1 the amount of wastage to be expected in small examples increases as $C/(1-r)$, and so a trade-off is involved. The best absolute worst case ratio is obtained by *FFS* (r) when $r \sim .622$, in which case we have $R_{FFS(r)} \sim 6.9863$.

The limitations inherent in the on-line approach are investigated by Brown, Baker and Katseff in [15]. They show that *any* on-line algorithm must obey $R_A \geqslant 2$. (The paper also contains bounds for on-line algorithms in the special case where they happen to be given sets of rectangles in some sorted order, but must still pack each item in turn, without being able to look ahead or to move an item once it is placed).

As a final contribution to strip packing we mention the results of Coffman and Gilbert [31] on *dynamic* packings in two dimensions. This is the natural extension of dynamic storage allocation in one dimension as defined in Section 3. In [31] the problem is specialized to squares and bottom up packings in a strip of width w. Extending the definitions in the one dimensional problem, let the squares in list L have a maximum size of $j \times j$ and assume that the total area of packed squares never exceeds mw. Let $BL(L)$ be the maximum height achieved by squares in L under the BOTTOM LEFT algorithm. For the asymptotic bound

$$C_{BL}^{\infty}(j) = \lim_{\substack{m \to \infty \\ w \to \infty}} \sup_L \frac{BL(L)}{m},$$

it is proved in [31] that

$$C_{BL}^{\infty}(j) \leqslant \frac{H_j}{\log 2 - \frac{1}{2}} \approx 5.177 \log j,$$

where H_j is the jth harmonic number. Moreover, lists are given which show that $C_{BL}^{\infty}(j) = \Theta(\log j)$. A number of results for finite m and w are given in [31] along with extensions to more than two dimensions.

There have been two papers to date that cover average case analysis for strip packing. In [42], Frederickson proposes an off-line algorithm combining *FFD* with specially tuned shelf sizes and specifically designed for the case when item sizes and widths are independently and uniformly distributed between O and C. Although the expected wastage may be large in absolute terms (proportional to $n^{3/4}$), the ratio of expected strip length to a lower bound on the optimal length (obtained by dividing the expected total area of rectangles by the strip width C) approaches 1 as n goes to infinity.

In [67], Hofri concentrates more on the on-line case, extending his earlier work with Coffman, So and Yao [32] on the expected behavior of one dimensional NEXT FIT to the strip packing problem. He considers two new on-line algorithms. The first is a level algorithm in which there is no initial reordering of the list of items, and hence the height of a level is not determined by its first rectangle but by the tallest, whichever one that might be. Otherwise the packing rule is basically a NEXT FIT one: an item is packed in the current level unless it cannot fit along the bottom, in which case it starts a new level, whose bottom is coincident with the top of the tallest item in the earlier level. Hofri calls this algorithm NEXT FIT, as opposed to NEXT FIT DECREASING HEIGHT where the items are preordered. Hofri's other new on-line algorithm is appropriate in the case where ninety degree rotations are allowed, and is called ROTATABLE NEXT FIT. This algorithm is the same as NEXT FIT except that each item is rotated before it is packed so that its height does not exceed its width.

Both of these two new on-line algorithms have $R_A^\infty = \infty$ and so are not very attractive from a worst case point of view. However, Hofri shows that when heights and widths are independent and uniformly distributed between O and C, they are not that much worse than NEXT FIT DECREASING HEIGHT, which has $R_{NFDH}^\infty = 2$ and is not an on-line algorithm. As $n \to \infty$, Hofri's results indicate that NEXT FIT DECREASING HEIGHT averages roughly 4/3 times the above-mentioned lower bound on optimal strip length. ROTATABLE NEXT FIT is only slightly worse, and NEXT FIT's ratio is only about 3/2.

Having introduced the case where ninety degree rotations are allowed, we should mention that some of the worst case results mentioned above also apply to this case, in that the values of R_A and

R_A^∞ are unchanged if such rotations are allowed in the construction of optimal packings. This holds true in particular for *NFDH* and *BLDW*, since the proofs of the bounds for these algorithms are based on pure area arguments. So far no algorithm has been found that attains improved guarantees by actually using such rotations itself, and the results mentioned above for the performance of strip packing algorithms when all items are squares (and hence ninety degree rotation cannot help) indicate that we can expect only limited improvements.

A rectangle packing problem closely related to strip packing is that of packing a given set of rectangles into an enclosing rectangle of minimum area. Strip packing is the special case where the width is fixed. In this general problem both length and width are allowed to vary. To date there has not been much work on this problem from a performance guarantee point of view. Two papers of interest have addressed the case when all the items to be packed are squares. In [81], Kleitman and Krieger show that a collection of squares whose total area is unity can always be packed into a rectangle with area $4/\sqrt{6}$, and this is the minimum area for which such a packing is guaranteed. Furthermore, a $2/\sqrt{3}$ by $\sqrt{2}$ rectangle is the unique rectangle that will always suffice. In [39], Erdös and Graham consider the minimum sized square required to contain a collection of unit squares, and show that this size can be nontrivially decreased if rotations other than ninety degrees are allowed.

Approximation algorithms for a problem complementary to these have been studied by Baker, Calderbank, Coffman and Lagarias [6]. Their problem was to pack the maximum number of squares from a given list into a rectangle of fixed dimensions. They analyzed both a BOTTOM UP INCREASING and a level-by-level NEXT FIT INCREASING algorithm and proved that $R_{BUI}^\infty = R_{NFI}^\infty = 4/3$, where again these are based on the inverted ratios $OPT(L)/A(L)$.

It must be pointed out that there is a sizable literature on square packing which we shall not survey here, primarily because it does not concern results closely related to approximation algorithms. The following list illustrates informally the variety of questions that have been asked: What is the smallest square sufficient to enclose n unit squares, rotations allowed? Can a rectangle with integer sides be tiled by a sequence of consecutive squares with sides 1,2,3,...? What is the smallest number of squares with integer sides into which a given square with integer sides can be partitioned? Extensive discussions of these and other intriguing but difficult questions can be found in [47,101], along with discussions of similar problems in packing circles and spheres.

The final rectangle packing problem we shall consider is a straightforward generalization of the one-dimensional case. Here the problem is once again simply to minimize the number of bins used, the bins now being large rectangles of some fixed dimensions into which the given set of rectangles must be packed. We first note that if the number of possible rectangle sizes is sufficiently small, a Gilmore-Gomory style linear programming approach can be applied [57] with useful results. For the general problem, the only algorithm which to date has been analyzed from the worst case point of view is a composite algorithm proposed by Chung, Garey and Johnson [20]. We shall denote this algorithm by "$FFDH \cdot FFD$," as it is based on the algorithm $FFDH$ for strip packing and FFD for one-dimensional bin packing. The idea of the algorithm is as follows: Suppose the standard bin has width W and height H. First use the $FFDH$ algorithm to pack the set of rectangles into a strip of width W. Next, decompose this packing into blocks corresponding to the levels created by $FFDH$. Each block can be viewed as a rectangle of width W and height the height of the level. Thus, packing these blocks into rectangular bins of width W becomes a simple one-dimensional bin packing problem, where the size of an item (block) is its height. Apply FFD to this one-dimensional problem.

The analysis of this algorithm in [20] shows that $2.022 \leqslant R^{\infty}_{FFDH \cdot FFD} \leqslant 2.125...$. Note that this leaves open the interesting possibility that $R^{\infty}_{FFDH \cdot FFD} = (R^{\infty}_{FFDH})(R^{\infty}_{FFD}) = (17/10)(11/9) = 2.0777...$, although the proof of such a result might well represent quite a challenge.

Quite recently Karp, Luby and Spaccamela [79] have considered a probability model of this problem under the assumption that the enclosing rectangles are unit squares. The dimensions of each of the n rectangles to be packed are independent random variables uniformly distributed over $[0,1]$. They devise an approximation algorithm requiring $n/4 + O(n^{1/2}\log n)$ bins on the average, thus generalizing the one-dimensional results in [76] and Frederickson's [42] strip packing results. Since the expected total area of the n rectangles is $n/4$, the ratio of the expected number of bins required by their algorithm to the expected number in a perfect packing approaches 1 as $n \rightarrow \infty$. Similar results are derived for extensions to more than two dimensions.

8. Directions for Future Research

In this section we briefly mention some of the open problems that we feel are significant from either a mathematical or practical point of view. First there is the basic problem of finding simpler and more general proof techniques. Although we have concentrated here on results rather than proof techniques, most of the results we have cited have only been proved by very problem-specific techniques that have rarely been exploited in analyzing related problems. It is true that researchers have been able to use intuition gained in studying the classical one-dimensional case in deriving results for the more complicated variants and generalizations, but unfortunately this is not often very apparent in the resulting proofs. The closest to a general method for proving results of this sort is the "weighting function" approach, as noted in Section 2, but so far the details of how this approach is used vary considerably from one problem to the next.

On a less fundamental level, there is of course the problem of finding better algorithms for the various problems, especially in the area of rectangle packing, and of tightening up the bounds on the algorithms already proposed but incompletely analyzed. There is also always room for new bin packing variants, the key being to find a variant that models practical problems *and* is susceptible to meaningful analysis. For instance, questions are often asked about the case when there are different types of bins (i.e., different sizes, different costs, etc.). The only work that can be cited here appears to be that of Langston [87] mentioned in Section 5.

Another technical problem is the very fundamental one of lower bounds. We have mentioned a number of lower bound results for on-line algorithms and we have cited the important recent developments in fully polynomial approximation schemes by Fernandez de la Vega and Lueker [40] and Karmarkar and Karp [74]. But the question of such schemes for two dimensional packings remains open, as does the question: Is there a polynomial time algorithm for the one dimensional problem that always comes within some additive constant of the optimum?

Our final remarks concern probabilistic analysis. We have noted the impressive increase in this research in the past few years. The most far-reaching contributions appear to have been those establishing rates of convergence to optimality of certain approximation algorithms, in the sense of expected performance. These results have usually been in the form of bounds that leave considerable room for future improvements. There is the added challenge, of course, to provide measures of second moments, and indeed, distributions of objective functions.

A characteristic weakness of many of the results to date has been that the algorithms analyzed have been chosen for their mathematical tractability rather than their attractiveness from a practical point of view. A prominent open problem is still a satisfactory average-case analysis of FIRST FIT, not to mention FIRST FIT DECREASING. A new such problem is to extend the $O(n^{-\log n})$ performance estimate to the more natural largest-pair-first set-differencing algorithm [75] introduced in Section 4. In view of the methods currently available, expecting exact results may be unreasonable. However, good bounds and results for interesting special cases would appear to be well worth the effort.

9. References

[1]   Albano, A. and Sapuppo, G., "Optimal allocation of two-dimensional irregular shapes using heuristic search methods," *IEEE Trans. Syste., Man, Cybern.*, SMC-10 (1980), 242-248.

[2]   Assmann, S. B., Doctoral Dissertation, Department of Mathematics, M.I.T., Cambridge, Mass. (1983).

[3]   Assmann, S. B., Johnson, D. J., Kleitman, D. J. and Leung, J. Y-T., "On a dual version of the one-dimensional bin packing problem," *J. of Algorithms* (to appear).

[4]   Baker, B. S., "A new proof for the first-fit decreasing bin-packing algorithm," Technical Memorandum (1983), Bell Laboratories, Murray Hill, N.J. 07974.

[5]   Baker, B. S., Brown, D. J., and Katseff, H. P., "A 5/4 algorithm for two-dimensional packing," *J. of Algorithms*, 2 (1981), 348-368.

[6]   Baker, B. S., Calderbank, A. R., Coffman, E. G., Jr., and Lagarias, J. C., "Approximation algorithms for maximizing the number of squares packing into a rectangle," *SIAM J. of Alg. Disc. Meth.* (to appear).

[7]   Baker, B. S. and Coffman, E. G., Jr., "A tight asymptotic bound for next-fit-decreasing bin-packing," *SIAM J. Alg. Disc. Meth.* 2 (1981), 147-152.

[8]   Baker, B. S., Coffman, E. G., Jr., and Rivest, R. L., "Orthogonal packings in two dimensions," *SIAM J. Comput.* 9 (1980), 846-855.

[9]   Baker, B. S. and Schwarz, J. S., "Shelf algorithms for two-dimensional packing problems," *SIAM J. Comput.* (to appear).

[10] Biro, M. and Boros, E., "A network flow approach to non-guillotine cutting problems," *Working Paper* MO/30 (1982), Computer and Automation Institute, Hungarian Academy of Sciences, Budapest.

[11] Brown, A. R., *Optimum Packing and Depletion*, American Elsevier, New York (1971).

[12] Brown, D. J., "A lower bound for on-line one-dimensional bin packing algorithms," Technical Report R-864 (1979), Coordinated Science Laboratory, University of Illinois, Urbana, IL.

[13] Brown, D. J., "An improved BL lower bound," *Inf. Proc. Letters* 11 (1980) 37-39.

[14] Brown, D. J., private communication (1980).

[15] Brown, D. J. and Baker, B. S. and Katseff, H. P., "Lower bounds for the on-line two-dimensional packing algorithms," *Acta Informatica*, 18 (1982), 207-225.

[16] Bruno, J. L. and Downey, P. J., "Probbilistic bounds on the performance of list scheduling," *Tech. Rep. TR 82-19*, Computer Science Dept., University of Arizona, Tucson, Ariz.

[17] Chandra, A. K. and Wong, C. K., "Worst-case analysis of a placement algorithm related to storage allocation," *SIAM J. Comput.* 4 (1975), 249-263.

[18] Chandra, A. K., Hirschberg, D. S., and Wong, C. K., "Bin packing with geometric constraints in computer network design," Computer Science Research Report RC 6895 (1977), IBM Research Center, Yorktown Heights, New York.

[19] Christofides, N. and Whitlock, C., "An algorithm for two-dimensional cutting problems," *Oper. Res.* 25 (1977), 30-44.

[20] Chung, F. R. K., Garey, M. R. and Johnson, D. J., "On packing two-dimensional bins," *SIAM J. Alg. Disc. Meth.* 3 (1982), 66-76.

[21] Cody, R. A. and Coffman, E. G., Jr., "Record allocation for minimizing expected retrieval costs on drum-like storage devices," *Journal of the ACM* 23 (1976), 103-115.

[22] Coffman, E. G., Jr., "An introduction to proof techniques for packing and sequencing algorithms," in *Deterministic and Stochastic Scheduling*, M.A.H. Dempster, et al. (eds.),

(1982), 245-270, Reidel Publishing Co., Amsterdam.

[23] Coffman, E. G., Jr., "An introduction to combinatorial models of dynamic storage allocation," *SIAM Review* (to appear).

[24] Coffman, E. G., Jr., Frederickson, G. and Lueker, G. S., "A note on expected makespans for largest-first sequences of independent tasks on two processors," *Math of OR* (to appear).

[25] Coffman, E. G., Jr., Frederickson, G. N. and Lueker, G. S., manuscript in preparation.

[26] Coffman, E. G., Jr., Garey, M. R., and Johnson, D. S., "An application of bin-packing to multiprocessor scheduling," *SIAM J. Comput.* 7 (1978), 1-17.

[27] Coffman, E. G., Jr., Garey, M. R., and Johnson, D. S., "Dynamic bin packing," *SIAM J. Comput.* 12 (1983), 227-258.

[28] Coffman, E. G., Jr., Garey, M. R. and Johnson, D. S., "Performance of packing algorithms for divisible sequences of item sizes," paper in preparation.

[29] Coffman, E. G., Jr., Garey, M. R., Johnson, D. S., and Tarjan, R. E., "Performance bounds for level-oriented two-dimensional packing algorithms," *SIAM J. Comput.* 9 (1980), 808-826.

[30] Coffman, E. G., Jr. and Gilbert, E. N., "On the expected relative performance of list scheduling," Technical Memorandum, Bell Laboratories, Murray Hill, N.J. 07974 (1983).

[31] Coffman, E. G., Jr. and Gilbert, E. N., "Dynamic first-fit packings in two or more dimensions," Technical Memorandum, Bell Laboratories, Murray Hill, N.J. 07974 (1983).

[32] Coffman, E. G., Jr., Hofri, M., So, K., and Yao, A. C., "A stochastic model of bin packing," *Inf. and Control* 44 (1980), 105-115.

[33] Coffman, E. G., Jr., and Leung, J. Y., "Combinatorial analysis of an efficient algorithm for

processor and storage allocation," *SIAM J. Comput.* 8 (1979), 202-217.

[34] Coffman, E. G., Jr., Leung, J. Y., and Ting, D. W., "Bin packing: maximizing the number of pieces packed," *Acta Informatica* 9 (1978), 263-271.

[35] Coffman, E. G., Jr. and Sethi, R., "A generalized bound on LPT sequencing," *RAIRO-Informatique* 10 (1976), 17-25.

[36] Dempster, M. A. H., Fisher, M. L., Jansen, L., Lageweg, B. J., Lenstra, J. K. and Rinnooy Kan, A. H. G., "Analysis of heuristics for stochastic programming: Results for Hierarchical scheduling problems," *Operations Res.,* 29 (1981), 707-716.

[37] Deuermeyer, B. L., Friesen, D. K. and Langston, M. A., "Maximizing the minimum processor finish time in a multiprocessor system," *SIAM J. Alg. Disc. Meth.* 3 (1982), 190-196.

[38] Easton, M. C. and Wong, C. K., "The effect of a capacity constraint on the minimal cost of a partition," *J. Assoc. Comput. Mach.* 22 (1975), 441-449.

[39] Erdös, P. and Graham, R. L., "On packing squares with equal squares," *J. Combinatorial Theory Ser. A* 19 (1975), 119-123.

[40] Fernandez de la Vega, W. and Lueker, G. S., "Bin packing can be solved within 1+ε in linear time," *Combinatorica* 1 (1981), 349-355.

[41] Finn, G. and Horowitz, E., "A linear time approximation algorithm for multiprocessor scheduling," *BIT* 19 (1979), 312-320.

[42] Frederickson, G. N., "Probabilistic analysis for simple one- and two-dimensional bin packing algorithms," *Inf. Proc. Letters* 11 (1980), 156-161.

[43] Frenk, H. and Rinnooy Kan, A. H. G., "The asymptotic optimality of the LPT heuristic," Erasmus University, Rotterdam, The Netherlands, (to be published).

[44] Friesen, D. K., "Sensitivity analysis for heuristic algorithms," Technical Report UIUCDCS-R-78-939 (1978), Dept. Comp. Sci., Univ. of Illinois, Urbana, IL. *SIAM J. Comput.*, (to appear).

[45] Friesen, D. K. and Langston, M. A., "Analysis of a compound bin-packing algorithm," (to appear).

[46] Galambos, G. and Turan, G., Laboratory of Cybernetics, Josef Attila University, Szeged, Hungary (private communication).

[47] Gardner, M., "Some packing problems that cannot be solved by sitting on the suitcase," in Mathematical Games column, *Scientific American,* Oct. 1979, 18-26.

[48] Garey, M. R. and Graham, R. L., "Bounds on multiprocessor scheduling with resource constraints," *SIAM J. Comput.* 4 (1974), 187-200.

[49] Garey, M. R., Graham, R. L., and Johnson, D. S., "On a number-theoretic bin packing conjecture," *Proc. 5th Hungarian Combinatorics Colloquium,* North-Holland, Amsterdam (1978), 377-392.

[50] Garey, M. R., Graham, R. L., Johnson, D. S. and Yao, A. C., "Resource constrained scheduling as generalized bin packing," *J. Combinatorial Theory Ser. A* 21 (1976), 257-298.

[51] Garey, M. R. and Johnson, D. S., "Approximation algorithms for combinatorial problems: an annotated bibliography," in J. F. Traub (ed.), *Algorithms and Complexity:* New Directions and Recent Results, Academic Press, New York (1976), 41-52.

[52] Garey, M. R. and Johnson, D. S., *Computers and Intractability: A Guide to the Theory of NP-Completeness,* W. H. Freeman and Co., San Francisco (1979).

[53] Garey, M. R. and Johnson, D. S., "Approximation algorithms for bin-packing problems — A survey," in *Analysis and Design of Algorithms in Combinatorial Optimization,* G. Ausiello

and M. Lucertini (eds.), Springer-Verlag, New York, 1981, 147-172.

[54] Garey, M. R. and Johnson, D. S., paper in preparation.

[55] Gilmore, P. C. and Gomory, R. E., "A linear programming approach to the cutting stock problem," *Operations Res.* 9 (1961), 849-859.

[56] Gilmore, P. C. and Gomory, R. E., "A linear programming approach to the cutting stock program — Part II," *Operations Res.* 11 (1963), 863-888.

[57] Gilmore, P. C. and Gomory, R. E., "Multistage cutting stock problems of two and more dimensions," *Operations Res.* 13 (1965), 94-120.

[58] Golan, I., "Two orthogonal oriented algorithms for packing in two dimension," Report 1979/311/MHM, Computer Center M.O.D., P. O. Box 2250, Haifa, Israel (1979).

[59] Golan, I., "Performance bounds for orthogonal, oriented two-dimensional packing algorithms," *SIAM J. Comput.* 10 (1981), 571-582.

[60] Graham, R. L., "Bounds for certain multiprocessing anomalies," *Bell System Tech. J.* 45 (1966), 1563-1581.

[61] Graham, R. L., "Bounds for Dynamic Storage Allocation Strategies," Technical Memorandum, Bell Laboratories, Murray Hill, N.J. (1968).

[62] Graham, R. L., "Bounds on multiprocessing timing anomalies," *SIAM J. Appl. Math.* 17 (1969), 263-269.

[63] Graham, R. L., "Bounds on multiprocessing anomalies and related packing algorithms," *Proc. 1972 Spring Joint Computer Conference,* AFIPS Press, Montvale, N.J. (1972), 205-217.

[64] Graham, R. L., "Bounds on performnce of scheduling algorithms," in E. G. Coffman, Jr. (ed.), *Computer and Job-Shop* Scheduling Theory, John Wiley & Sons, New York (1976), 165-227.

[65] Graham, R. L., Lawler, E. L., Lenstra, J. K., and Rinnooy Kan, A. H. G., "Optimization and approximation in deterministic sequencing and scheduling: a survey," *Annals Disc. Math.* 5 (1979), 287-326.

[66] Hoffman, U., "A class of simple online bin packing algorithms," *Computing,* 29 (1982), 227-239.

[67] Hofri, M., "Two dimensional packing: expected performance of simple level algorithms," *Inf. and Control* 45 (1980), 1-17.

[68] Hofri, M., "Bin-packing: An analysis of the Next-Fit algorithm," *Tech. Rep. No. 242,* Dept. of Computer Science, The Technion, Haifa, Israel (1982).

[69] Johnson, D. S., "Near-optimal bin packing algorithms," Technical Report MAC TR-109 (1973), Project MAC, Masschusetts Institute of Technology, Cambridge, Mass.

[70] Johnson, D. S., "Fast algorithms for bin packing," *J. Comput. Syst. Sci.* 8 (1974), 272-314.

[71] Johnson, D. S., "The *NP*-completeness column: An ongoing guide, *J. of Algorithms* 2 (1981), 393-405 (and succeeding issues).

[72] Johnson, D. S., Demers, A., Ullman, J. D., Garey, M. R., and Graham, R. L., "Worst-case performance bounds for simple one-dimensional packing algorithms," *SIAM J. Comput.* 3 (1974), 299-325.

[73] Karmarkar, N., "Probabilistic analysis of some bin-packing problems," *Proc. 23rd Ann. Symp.* on Foundations of Computer Science, IEEE Computer Soc., Nov. 1982 (full paper to appear elsewhere).

[74] Karmarkar, N. and Karp, R. M., "An efficient approximation scheme for the one-dimensional bin packing problem," *Proc. 23rd Ann. Symp. on Foundations of Computer Science,* IEEE Computer Soc., Nov. 1982 (full paper to appear elsewhere).

[75] Karmarkar, N. and Karp, R. M., "The differencing method of set partitioning," Computer Science Div., University of California, Berkeley, Calif., to be published.

[76] Karmarkar, N., Karp, R. M., Lueker, G. S. and Murgolo, F., Computer Science Div., University of California, Berkeley, Calif., paper in preparation.

[77] Karmarkar, N., Karp, R. M., Lueker, G. S., and Odlyzko, A., Bell Laboratories, Murray Hill, N.J., paper in preparation.

[78] Karp, R. M., "Reducibility among combinatorial problems," in R. E. Miller and J. W. Thatcher (eds.), *Complexity of Computer Computations,* Plenum Press, New York (1972), 85-103.

[79] Karp, R. M., Luby, M. G. and Spaccamela, A. M., "Probabilistic analysis of multidimensional bin packing problems," Computer Science Div., University of California, Berkeley, Calif., paper in preparation.

[80] Kaufman, M. T., "An almost-optimal algorithm for the assembly line scheduling problem," *IEEE Trans. Computers* C-23 (1974), 1169-1174.

[81] Kleitman, D. J. and Krieger, M. K., "An optimal bound for two dimensional bin packing," *Proc. 16th Ann. Symp. on Foundations of Computer Science,* IEEE Computer Society, Long Beach, CA (1975), 163-168.

[82] Knödel, W., "A bin-packing algorithm with complexity $O(n \log n)$ and performance 1 in the stochastic limit," *Proc., 10th Symp. on Math. Foundations in Comp. Sci.* (1981). (to appear in *Lecture Notes in Computer Science,* Springer-Verlag).

[83] Knuth, D. E., *Fundamental Algorithms,* Vol. 1, Second edition, Addison-Wesley (1973).

[84] Kou, L. T. and Markowsky, G., "Multidimensional bin packing algorithms," *IBM J. Res. & Dev.* 21 (1977), 443-448.

[85]   Krause, K. L., Shen, Y. Y., and Schwetman, H. D., "Analysis of several task-scheduling algorithms for a model of multiprogramming computer systems," *J. Assoc. Comput. Mach.* **22** (1975), 522-550.

[86]   Langston, M. A., *Processor Scheduling with Improved Heuristic Algorithms,* Doctoral dissertation, Texas A&M University, College Station, Texas (1981).

[87]   Langston, M. A., "Performance of bin-packing heuristics for maximizing the number of pieces packed into bins of different sizes," *Tech. Rep. No. CS-82-090,* Computer Science Dept., Washington State University, Pullman, Wash. (1982).

[88]   Langston, M. A., "Improved LPT scheduling for identical processor systems," *RAIRO-Technique et Science Informatiques,* 1 (1982), 69-75.

[89]   Langston, M. A., Improved 0/1 interchange scheduling," *BIT* 22 (1982), 282-290.

[90]   Lee, C. C. and Lee, D. T., "A simple on-line packing algorithm," Dept. of Electrical Engineering and Computer Science (1983), Northwestern Univ., Evanston, Ill. 60201 (to appear).

[91]   Liang, F. M., "A lower bound for on-line bin packing," *Information Processing Lett.* **10** (1980), 76-79.

[92]   Loulou, R., "Tight bounds and probabilistic analysis of two heuristics for parallel processor scheduling," *Tech. Rep.,* Faculty of Management, McGill University, Montreal (1982).

[93]   Loulou, R., "Probabilistic behavior of optimal bin packing solutions," *Tech. Rep.,* Faculty of Management, McGill University, Montreal (1982).

[94]   Lueker, G. S., "An average-case analysis of bin packing with uniformly distributed item sizes," *Tech. Rep. No. 181* (1982), Dept. of Information and Computer Science, University of California, Irvine, CA 92717.

[95] Lueker, G. S., "Bin packing with items uniformly distributed over intervals $[a, b]$," Dept. of Information and Computer Science (1983), University of California, Irvine, CA 92717 (to be published).

[96] Maruyama, K., Chang, S. K., and Tang, D. T., "A general packing algorithm for multidimensional resource requirements," *Internat. J. Comput. Infor. Sci.* **6** (1977), 131-149.

[97] Ong, H. L., Magazine, M. J., and Wee, T. S., "Probabilistic analysis of bin-packing heuristics," *Operations Res.* (to appear).

[98] Robson, J. M., "Bounds for some functions concerning dynamic storage allocation," *Journal of the ACM* 21 (1974), 491-499.

[99] Robson, J. M., "Worst-case fragmentation of first-fit and best-fit storage allocation strategies," *Computer J.,* 20 (1977), 242-244.

[100] Sahni, S., "Algorithms for scheduling independent tasks," *Journal of the ACM* 23 (1976), 116-127.

[101] Schrijver, A. (ed.), *Packing and Covering in Combinatorics,* published by Mathematical Centre, Tweede Boerhaavestraat 49, Amsterdam (1979).

[102] Shapiro, S. D., "Performance of heuristic bin packing algorithms with segments of random length," *Information and Control* 35 (1977), 146-148.

[103] Shearer, J. B., "A counterexample to a bin packing conjecture, *SIAM J. Alg. Disc. Meth.* 2 (1981), 309-310.

[104] Sleator, D. K. D. B., "A 2.5 times optimal algorithm for bin packing in two dimensions," *Information Processing Lett.* **10** (1980), 37-40.

[105] Steele, M., Stanford University (private communication).

[106] Taylor, D. B., "Container stacking: an application of mathematical programming," Draft (1979).

[107] Wee, T. S. and Magazine, M. J., "Assembly line balancing as generalized bin-packing," *Operation Res. Letters,* 1 (1982), 56-58.

[108] Wong, C. K. and Yao, A. C., "A combinatorial optimization problem related to data set allocation," *Rev. Francaise Automat. Informat. Recherche Operationelle Ser. Bleue* **10.5** (suppl.) (1976), 83-95.

[109] Yao, A. C., "New algorithms for bin packing," *J. Assoc. Comput. Mach.* **27** (1980), 207-227.

[110] Yue, P. C. and Wong, C. K., "On the optimality of the probability ranking scheme in storage applications," *J. Assoc. Comput. Mach.* **20** (1973), 624-633.

NETWORK DESIGN WITH NON SIMULTANEOUS FLOWS

Mario Lucertini
Istituto di Analisi dei Sistemi ed Informatica
del CNR and
Dipartimento di Informatica e Sistemistica,
Università di Roma, Viale Manzoni 30, 00185 Roma - Italy

Giuseppe Paletta
Consorzio per la Ricerca e le Applicazioni di Informatica
Via Bernini 5, 87036 Rende (CS) - Italy

ABSTRACT

The problem of finding the minimum total cost edge capacities, such that all demand vectors in a given set are non simultaneously satisfied, is analyzed both for directed and non directed networks. In both cases the optimal solution is shown to be the sum of basic networks with suitable uniform edge capacities that can be obtained with standard shortest path algorithms.

Work performed by CRAI under contract n. 82.00024.73 "Progetto Finalizzato Trasporti".

1. INTRODUCTION

The network synthesis problem with non simultaneous flow requirements has only recently become a research subject in mathematical programming (MI). Nevertheless this subject has been widely recognized as a main topic in distributed capacity expansion problems (FR).

Many classical results of network flow theory (HU) can be directly utilized in the non simultaneous flow environment, but new problems arise and new approaches must be pointed out.

In this paper the single commodity network design problem with non simultaneous flow requirements is analyzed. More precisely, the problem formally stated in section 2 and solved in the following is the problem of finding the minimum total cost edge capacities, such that all demand vectors in a given set are not simultaneously satisfied. Such problem is solved both for directed and non directed networks.

The basis ideas behind the model utilized have been first introduced in (LU) and (LP1).

2. MODEL FORMULATION

Let $G(N,A,c)$ be a network where N is the set of nodes (a source s, two or more sinks $1,2,\ldots p$, $n-1-p$ intermediate nodes, let I be the corresponding set), A is a set of edges $|A| = m$ and c is a capacity m-vector with non negative entries. Let $g(i,j)$, $(i,j) \in A$, be the edge capacity expansion unitary costs and R the set of demand p-vectors $d(d(i)$ is the amount of flow required in node $i = 1,2,\ldots,p)$.

In order to have simpler formulations of the results and simpler notations, we analyze, in sections 3, 4 and 5, a two sinks network, a finite set $R(|R| = r)$ and a unique source. Almost all the results can be easily extended to a network with more sources or sinks (provided that the problem will be

single commodity) and a set R with infinite elements like,
for example, a (bounded) polyhedron.The graph can be either
directed or undirected (in this case $f(i,j) + f(j,i) \leq c(i,j)$
and $f(i,j) \geq 0$ $\forall$ (i,j) where $f(i,j)$ is the flow on
$(i,j) \in A$ from i to j and $c(i,j)$ is the capacity of (i,j),
obviously in practice we can suppose that $f(i,j) \cdot (f(j,i) = 0)$.
If in a statement it is not specified whether or not the
graph is directed, it meas that the statement holds both for
directed and undirected graphs (graphs with both directed and
undirected edges are not considered).

Capacity expansion problem (CE)

*Find the minimum total cost edge capacities such that
all demand vectors in R are non simultaneously satisfied (i.e.
$G(N,A,c)$, with an infinite capacity source, which can satisfy
either d^1 or d^2 or d^3....or d^r, with $d^i \in R$ and*

$\sum_{(i,j) \in A} c(i,j) \cdot g(i,j)$ *is minimum).*

The problem of finding the set of flow vectors feasible
for $G(N,A,c)$ is analyzed in section 3 and it is shown that
for each network there exists a "flow equivalent" network
equal to the "sum of basic networks" with uniform edge capaci
ties (the terms "flow equivalent" and "sum of basic networks"
will be defined in section 3).

In section 4 and 5 the solution of CE is given for direc
ted and non directed networks respectively. In both cases the
optimal solution is shown to be the sum of three basic networks with suitable uniform edge capacities. In the first case
the three basic networks are two paths from s to 1 and 2
respectively and a tree with endpoints in s, 1 and 2. In the
second case the three basic networks are all paths (from s to
1, from s to 2 and from 1 to 2). In section 6 the extension
to multiterminal networks is presented.

3. NETWORK ANALYSIS

Let be given $G(N,A,c)$. Let CT12, CT1 and CT2 be the capacities of the minimal cuts $(X,\bar{X})$ such that $(s \in X$ and $1,2 \in \bar{X})$, $(s \in X, 1 \in \bar{X})$ and $(s \in X, 2 \in \bar{X})$ respectively (remark that CT1 $\leq$ CT12, CT2 $\leq$ CT12 and CT1 + CT2 $\geq$ CT12). Let q be a 2-vector where $q(1)$ and $q(2)$ denote the flow entering sink 1 and sink 2 respectively and Q the set of sink flow vectors q such that both the capacity constraints and the flow conservation constraints are satisfied (what goes out of node i must be equal to what comes in for all intermediate nodes $i \in I$); we call q feasible if $q \in Q$.

THEOREM 1 (HU) - Given $G(N,A,c)$, q is feasible if and only if:

$$q(1) \leq CT1, \quad q(2) \leq CT2, \quad q(1) + q(2) \leq CT12$$

Obviously the max flow from s to 1 is CT1, from s to 2 is CT2 and from s to both 1 and 2 is CT12; in general the flow from s to 1 depends on the flow from s to 2 and viceversa.

Let G1, G2, G3 be three networks with the same nodes and edges but with different capacity vectors $c(G1)$, $c(G2)$, $c(G3)$.

DEFINITION 1 *(Sum of networks) - G3 is said to be the sum of G1 and G2 if $c(G3) = c(G1) + c(G2)$.*

DEFINITION 2 *(Flow equivalent networks) - G1 and G2 are said to be flow equivalent if $Q(G1) = Q(G2)$ (where $Q(G)$ indicates the feasible set of sink flows of G).*

LEMMA 1 - *G1 and G2 are flow equivalent if and only if they have the same capacities of minimum cuts CT1, CT2 and CT12.*

DEFINITION 3 *(Basic networks) - Given $G(N,A,c)$, a basic network is any network $GB(N,A,cB)$ such that all entries of capacity vector cB are either 0 or $c' \in \mathbb{R}^{+}$ and all the edges*

*with nonzero uniform capacity form one of the following sub-
networks:*

1) *path from* s *to* 1

2) *path from* s *to* 2

3) - *(oriented network) tree with root in* s *and leaves
   in* 1 *and* 2 *and a unique branching node;*

    - *(non-oriented network) cycle connecting* s, 1 *and* 2
   (also sum of paths from s *to* 1, *from* s *to* 2 *and
   from* 1 *to* 2 *with the same uniform capacity).*

Let GB1, GB2 and GB3 be the set of basic networks satisfying condition 1, 2 and 3 (definition 3) respectively.

THEOREM 2 - For each network $G(N,A,c)$, there exists a
flow equivalent network $GE(N,A,cE)$ with $cE \leq c$, sum of basic
networks. □

In the following of this section this result will be
proved and some useful notations for the synthesis will be
introduced.

Remark that the proof of theorem 2 without the constraints $cE \leq c$ is trivial. In fact, given $G(N,A,c)$ and the
minimal cuts CT1, CT2 and CT12, we can easily build an equiva
lent network, sum of three networks G1, G2 and G3, each of
them sum of basic networks in the set GB1, GB2 and GB3 respectively, with minimal cuts given by:

$$C1T1 = C1T12 = CT12 - CT2, \ C1T2 = 0$$

$$C2T1 = 0, \ C2T2 = C2T12 = CT12 - CT1 \qquad (1)$$

$$C3T1 = C3T2 = C3T12 = CT1 + CT2 - CT12$$

where CiT1, CiT2 and CiT12 are the minimal cuts of the network Gi, $i = 1,2,3$.

DEFINITION 4 *(Maximal flow) - Given* $G(N,A,c)$, *a sink
flow vector* q *is said to be maximal if it is not possible to*

augment $q(1)$ *without reducing* $q(2)$ *and viceversa.*

LEMMA 2 - *Given* $G(N,A,c)$, *a maximal sink flow vector* q *satisfies the following constraints:*

$$CT12 - CT2 \leq q(1) \leq CT1$$
$$CT12 - CT1 \leq q(2) \leq CT2$$
$$q(1) + q(2) = CT12$$

Remark that G1 and G2 satisfy the minimum maximal flow on sink 1 and sink 2 respectively; G3 satisfies the flow (CT1 + CT2 - CT12) that can be sent from s to 1 or to 2, but not simultaneously.

The way of proving the existence of a flow equivalent network with cE $\leq$ c is conceptually simple but quite long. In the following we give an outline of the proof.

It is simple to find G1 and G2, sum of basic networks belonging to GB1 and GB2 respectively, such that the sum satisfies the minimum maximal flow from s to 1 and from s to 2 simultaneously, and c(G1 + G2) $\leq$ c; G1 and G2 can be obtained by solving a min-cost-max-flow problem, with demand on the sinks given by (CT12 - CT2) and (CT12 - CT1) and capacity constraints c. In fact the two flows are independent.

(G1 + G2) has minimal cuts CST1, CST2 and CST12 given by:

$$CST1 = CT12 - CT2, \quad CST2 = CT12 - CT1,$$
$$CST12 = 2 \cdot CT12 - CT1 - CT2;$$

since G has cuts CT1, CT2 and CT12, we obtain:

$$CT1 - CST1 = CT2 - CST2 = CT12 - CST12 = CT1 + CT2 - CT12$$

hence there exists a network $\overline{G3}$ with all minimal cuts equal to $(CT1 + CT2 - CT12)$, obtained as:

$$c(\overline{G3}) = c(G) - c(G1 + G2).$$

It is possible to verify that $\overline{G3}$ is flow equivalent to a network G3, sum of basic networks belonging to GB3.

4. DIRECTED NETWORK SYNTHESIS

In order to solve CE it is not necessary to take into account the whole region R, but only few parameters characterizing the region. More precisely, let $\hat{d}$ be a demand vector such that the sum of the entries wil be maximum (i.e. $dM12 = \sum_i (\hat{d}(i) = \max_{d \in R} \sum_i d(i))$ and dMi be the maximum value of demand vectors entry i (i.e. $dMi = \max_{d \in R} d(i)$).

LEMMA 3 - *The demand vectors* d ∈ R *are non simultaneously satisfied if and only if:*

$$CT1 \geq dM1, \quad CT2 \geq dM2, \quad CT12 \geq dM12.$$

PROOF - If the minimal cuts satisfy the inequalities then all d ∈ R for theorem 1 can be satisfied. On the other hand if all d ∈ R can be satisfied, the minimal cuts are at least dM1, dM2 and dM12.

Capacity expansion for directed network (CED)

Find the minimum total cost edge capacities such that the minimal cuts C1T1, C2T2 *and* C2T12 *(characterizing the equivalent networks* G1, G2 *and* G3*) will be non negative and satisfy the constraints:*

$$CT1 = C1T1 + C3T12 \geq dM1$$

$$CT2 = C2T2 + C3T12 \geq dM2 \qquad (2)$$

$$CT12 = C1T1 + C2T2 + C3T12 \geq dM12$$

In fact, in the following, we will prove that the first two inequalities are satisfied (in an optimal solution) as equalities and CT12 can be either equal to dM12 or equal to dM1 + dM2.

Let P1 (P2) be the shortest path from s to 1(2) with weights $g(i,j)$ and h1(h2) be the corresponding lenght; let Q12 be the tree with root in s and leaves in 1 and 2 (and a unique branching node), with minimum total weight h12 (i.e. $h12 = \min_{k \in (N-s)}(h(s,k)+h(k,1)+h(k,2))$ where $h(i,j)$ indicates the lenght of the shortest path from i to j). h1(h2) units of money invested on P1(P2) are the cheapest way to send a unit of flow from s to 1(2); h12 units of money invested on Q12 is the cheapest way to send a unitary flow from s to 1 or to 2 but not simultaneously. When we refer to investment on paths or on trees we intend uniform investment, i.e. such that the capacity of all edges of the path or the tree be the same. Let GP1(c'), GP2(c') and GQ12(c') be the networks obtained by assigning the capacity c' to all edges of the path P1, P2 and the tree Q12 respectively (and the capacity zero to all other edges). Remark that GP1(c') $\in$ GB1, GP2(c') $\in$ GB2 and GQ12(c') $\in$ GB3 for all c' $\in \mathbf{R}^+$; furthermore GP1(C1T1) (GP2(C2T2)) is the cheapest way to obtain a minimal cut C1T1(C2T2), and GQ12(C3T12) is the cheapest way to obtain minimal cuts equal to C3T12 between s and 1, s and 2, 1 and 2. We have in fact outlined the proof of the following result.

THEOREM 3. There exist three non negative numbers c', c" and c"' such that one optimal solution of CED can be ob-

tained as the sum of basic networks GP(c'),GP(c") and GQ12(c''')
(i.e. the optimal solution can be obtained by investing only
on the paths P1, P2 and on the tree Q12). □

The problem now reduces itself to finding C1T1, C2T2
and C3T12 such that the demand constraints will be satisfied
and the total cost will be minimized.

From such values the edge capacities can be easily
calculated. Remark that h12 $\leq$ h1 + h2.

THEOREM 4. A capacity vector, optimal for CED, satisfies
the following relations. If h12 < h1 + h2 then:

C1T1 = dM12-dM2, C2T2 = dM12-dM1, C3T12 = dM1+dM2-dM12.

If h12 = h1+h2 then an optimal solution is given by:

C1T1 = dM1, C2T2 = dM2, C3T12 = 0.

PROOF. CED can be written as a linear programming problem with 3 variables and 3 inequality constraints:

min(h1·C1T1+h2·C2T2+h12·C3T12)

subject to the constraints (2). Solving this problem, we obtain the results of theorem 4. □

Remark that if h12 = h1+h2 but Q12 is not given as the
sum of P1 and P2 (with suitable capacities), then all capacity vectors are optimal if:

0 $\leq$ C3T12 < dM1+dM2-dM12, C1T1 = dM1-C3T12, C2T2 = dM2-C3T12

The optimal network capacities are given by:

c(G) = c(GP1(C1T1))+c(GP2(C2T2))+c(GQ12(C3T12)).   (3)

5. NON DIRECTED NETWORK SYNTHESIS

For nondirected networks similar results to the ones presented in section 4 hold. In particular lemma 3 holds and the expansion capacity problem can be written as:

Capacity expansion for non directed networks (CEN)

Find the minimum cost edge capacities such that the minimal cuts C1T1, C2T2 and C3T12 be non negative and satisfy the constraints:

$$C1T1 + 2 \cdot C3T12 \geq dM1$$
$$C2T2 + 2 \cdot C3T12 \geq dM2 \tag{4}$$
$$C1T1 + C2T2 + 2 \cdot C3T12 \geq dM12$$

Also in this case, in an optimal solution, the first two inequalities are satisfied as equalities and CT12=C1T1+C2T2+ 2·C3T12 can be either equal to dM12 or equal to dM1+dM2.

Let P1,P2,h1 and h2 be defined as in section4, Q12 be the minimum cost cycle connecting s, 1 and 2 and h12 be the corresponding lenght. Q12 in this case is obtained as the link of three shortest paths: P1,P2 and the shortest path P12 from 1 to 2 (or equivalently from 2 to 1), hence h12 = = h1+h2+h(12) where h(12)indicates the lenght of P12. Let GP1(c'), GP2(c') and GQ12(c') be the networks obtained by assigning the capacity c' to all edges of P1, P2 and Q12 respectively (and the capacity zero to all other edges).With these definitions theorem 3 holds also for non directed networks. Theorem 4 becames (remark that h12≤2(h1+h2)):

THEOREM 5. A capacity vector optimal for CEN satisfies the following relations: if h12 < 2(h1+h2) then:

$$C1T1 = (-dM2+dM12), \quad C2T2 = (-dM1+dM12)$$

$$C3T12 = (dM1+dM2-dM12)/2.$$

If h12 = (h1+h2) then an optimal solution is given by:

$$C1T1 = dM1, \quad C2T2 = dM2, \quad C3T12 = 0$$
□

Remark that, if h12 = 2(h1+h2) but Q12 is not given as the sum of P1 and P2 (with suitable capacities), then all capacity vectors are optimal if:

$$0 \le C3T12 \le (dM1+dM2-dM12)/2, C1T1 = dM1-C3T12,$$

$$C2T2 = dM2 - C3T12$$

The proof follows the same lines of the proof of Theorem 4; the resulting capacities c(G) are given by (3).

It is interesting to note that the difference core between the directed and the non directed network can be summarized in the following result. Let be given two non oriented networks G' and G", where G' is a minimum cost tree with endpoints s, 1 and 2, a unique branching node i ∈ I and uniform edge capacities c'; G" is the cycle obtained as the link of P1, P2 and P12 with uniform edge capacities c'; let h(G'(c')) and h(G"(c')) be the total costs of G'(c') and G"(c') respectively.

COROLLARY. G'(c') and G"(c'/2) are flow equivalent networks and

$$h(G'(c')) \ge h(G"(c'/2)).$$
□

6. EXTENSION TO MULTITERMINAL NETWORKS

All the results presented in the previous sections hold for multiterminal networks. In fact the single commodity case analyzed in this paper can allways be formulated as a unique

source problem by introducing a supersource connected to all
sources by infinite capacity edges. The problem is to deal
with the case of three or more sinks.

Both the cases of directed and non directed networks
can be analyzed in the same way, in the following we develop
the approach for the non directed networks.

Let S be the set of sinks and let us denote by $P(T)$ the
set of all the nonempty subsets of the set $T(|P(T)| = 2^{|T|}-1)$
and $P_o(T) = P(T)-T$. Let $c(T)$ (with $T \in P(S)$) be the (uniform)
capacity of the (basic) minimum cost network connecting the
source s and all the sinks belonging to the set T; in practice $c(i)$ $i \in S$ will be the capacity of the shortest path
from s to the sink i, $c(T)$ with $T \in P(S)$ and $|T| \geq 2$ will be
the capacity of the minimum lenght cycle connecting s and all
$i \in T$; let $h(T)$ be the corresponding lenght of the cycle or
the lenght of the shortest path from s to i in case $T = \{i\}$.

It is easy to verify that also in the multiterminal case
there always exists an optimal solution, sum of optimal basic
networks. Unfortunately the number of optimal basic networks
grows exponentially with the number of sinks.

Let $\delta(T)$ be a scalar obtained as the maximum value of
the sum of the demand vectors entries corresponding to the
set T (i.e. $\delta(T) = \max\limits_{d \in R} \sum\limits_{i \in T} d(i)$) and $\phi(T)$ be the set of all
subsets U of S such that $U \cap T \neq \emptyset$ and $|U| \geq 2$.

The problem can be formulated as a linear program as
follows:

$$\min z = \sum_{T \in P(S)} h(T)c(T)$$

$$\sum_{i \in T} c(i) + \sum_{U \in \phi(T)} 2 \cdot c(U) \geq \delta(T) \qquad \forall T \in P(S) \qquad (5)$$

LEMMA. *The solution of the* $(2^{|S|} - 1)$ *linear equations*

$$\sum_{i \in T} c(i) + \sum_{U \in \phi(T)} 2 \cdot c(U) = \delta(T) \quad (\forall T \in P(S)) \; is:$$

$$c(i) = \delta(S) - \delta(S-i) \qquad\qquad \forall i \in S$$

$$c(T) = \frac{1}{2}(\delta(S) - \delta(S-T) - \sum_{U \in P_o(T)} c(U)) \qquad \forall T \in P(S)$$
□

Lemma 4 allows the computation of all c(T) for increasing sizes of T provided that all inequalities in (5) will be satisfied as equalities.

Remark that, although the solution given by Lemma 4 can contain negative values of c(T), the resulting arc capacities (sum of the capacities of all basic networks utilizing the given arc) are allways nonnegative.

The problem now reduces to find under which hypotheses Lemma 4 provides the optimal solution of problem (5) and how to get the optimal solution if the hypotheses are not verified.

Let $\pi(T)$ be a partition in nonempty subsets of T and $\Pi(T)$ the set of all possible partitions of T.

THEOREM 6. If all minimum lenght cycles are uniques then all inequalities in (5) are satisfied in the optimal solution as equalities if and only if

$$c(T) < \sum_{R \in \pi(T)} c(R) \qquad \forall \pi(T) \in \Pi(T)$$

$$\forall T \subseteq S$$
□

Theorem 6 means that the solution given by Lemma 4 solves problem 5, if all the shortest lenght cycles cannot be obtained as sum of shortest lenght cycles over a smaller set of sinks.

The proof can be easily obtained by considering that if the statements of theorem 6 are not true, then there exists at least a cycle passing through the source two or more times; hence the total capacity leaving the source is greater than the minimum requirement (in fact we have in this case a positive flow entering in the source).

If some minimum lenght cycles are not unique we have an infinite optimal solution, see the remarks after theorem 4 and 5 for the two-dimensional case.

If the statements of theorem 6 are not true, then let ψ be the set of all cycles sum of cycles over a smaller set of sinks.

In this case the optimal solution of problem (5) can be simply obtained by deleting the constraints corresponding to the cycles in ψ and the columns corresponding to the capacities of the cycles in ψ. We obtain a reduced size problem of the same kind of the original one. As far as the complexity is concerned, the exponentiality of the algorithm with respect to the number of sinks depends on two different reasons: the exponential grow of the number of cycles and the computation of the minimum lenght cycles equivalent to a Travelling Salesman Problem with triangle inequality (ΔTSP [PS,RSL]).

7. INTEGRALITY CONSTRAINTS

In many applications the investments on the network edges cannot be chosen arbitrarily, but must belong to preassigned discrete set of values. The corresponding formulation is an integer programming problem.

It is interesting to observe that the optimal solutions of section 4 are integer, provided that the demand vectors are integers. The optimal solutions of section 5 (non direct-

ed networks) can be in some cases non integers. In fact, if
h12 < 2(h1+h2), the solution is integer only if (dM1+dM2-dM12)
is even.

If (dM1+dM2-dM12) is odd, the solution given in section
5 is not integer; but an integer solution can be easily found.
Remark that the capacities assigned directly to P1 and P2 are
integers and only the capacities assigned to Q12 are non in-
tegers(but multiple of 1/2); the total capacities assigned
directly or through Q12 to P1 and P2 are non integers.

The optimal integer solution is the best one among the
following three possibilities:

a) increase the capacities of P1 and P2 to the upper integer
   and decrease the capacities of P12 to the lower integer;
b) increase the capacities of P1 and P12 to the upper integer
   and decrease the capacities of P2 to the lower integer;
c) increase the capacities of P2 and P12 to the upper integer
   and decrease the capacities of P1 to the lower integer.

In the multiterminal network of section 6, the situation
changes: the solution can be non integer (but multiple of
1/2) but in general it cannot be found by choosing the best
one among all possible alternatives obtained by rounding the
non integer path capacities to the nearest integers.

In fact in this case new paths can become active in the
optimal solution (see figure).

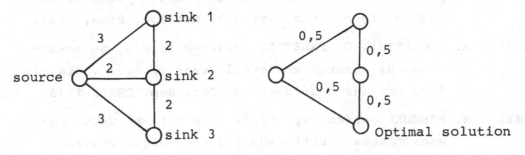

$$d^1 = \begin{pmatrix} 1 \\ 0 \\ 0 \end{pmatrix}$$

$$d^2 = \begin{pmatrix} 0 \\ 1 \\ 0 \end{pmatrix}$$

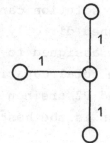

Optimal integer solution

$$d^3 = \begin{pmatrix} 0 \\ 0 \\ 1 \end{pmatrix}$$

REFERENCES

(FR)   J. FREIDENFELDS: *Capacity Expansion*, North Holland, 1981.

(HU)   T.C. HU: *Combinatorial algorithms*, Addison Wesley, 1982.

(LU)   M. LUCERTINI: *Bounded rationality in long term planning: a linear programming approach*, Metroeconomica, 1982.

(LP1)   M. LUCERTINI, G. PALETTA: *A class of network design problems with multiple demand: model formulation and an algorithmic approach*, Netflow, Pisa, 1983.

LP2)   M. LUCERTINI, G. PALETTA: *Progetto di reti in condizione di domande non simultanee: il caso a singolo bene con due nodi domanda*, Tec. Rep. CRAI, 1983.

(MI)   M. MINOUX: *Optimum synthesis of a network with non-simultaneous multicommodity flow requirements*, Ann. Dis. Math., Vol. 11, 1981.

(PS)   C. PAPADIMITRIOU, L. STEIGLITZ: *Some complexity re-
        sults for the TSP*, 8th ACM-STOC, 1976.

(RSL)   D. ROSENKRANTZ, R. STEARNS, P. LEWIS: *An analysis of
        general heuristics for the TSP*, SIAM J. Comp.,
        1977.

MINIMAL REPRESENTATIONS OF DIRECTED
HYPERGRAPHS AND THEIR APPLICATION TO
DATABASE DESIGN*

G. Ausiello, A. D'Atri
Università "La Sapienza" Roma, Italy

and

D. Saccà
CRAI, Via Bernini 5, 87030 Rende, Italy

ABSTRACT

In this paper the problem of minimal representations for sets of functional dependencies for relational databases is analyzed in terms of directed hypergraphs. Various concepts of minimal representations of directed hypergraphs are introduced as extensions to the concepts of transitive reduction and minimum equivalent graph of directed graphs. In particular we consider coverings which are the minimal representations with respect to all parameters which may be adopted to characterize a given hypergraph (number of hyperarcs, number of adjacency lists required for the representation, length of the overall description, etc.). The relationships among the various minimal coverings are discussed and the computational properties are analyzed. In order to derive such results a graphic representation of hypergraphs is introduced. Applications of these results to functional dependency manipulation are finally presented.

---

* This research has been partially supported by MPI Nat. Proj. on "Theory of algorithms".

1. INTRODUCTION

Hypergraphs are a generalization of the concept of graph [4] which have been extensively used for representing structures and concepts in several areas of computer science (see, for example [3,5,6,8,12,16]).

In this paper we consider a particular class of directed hypergraphs, the R-triangular hypergraphs which are a simple generalization of directed graphs.

In several applications of R-triangular hypergraphs, analogously to what happens in the case of graphs, the following concepts assume an important role: the concept of path (i.e. edge connection leading from a set of nodes to a single node), the concepts of closure (i.e. representation of all paths over a hypergraph), the concept of "minimal" covering (i.e. representation of the closure which is minimal under some respect).

In this paper R-triangular hypergraphs are applied for representing a set of functional dependencies among attributes in a relational database schema. In database design [10],[15] a major role is played by functional dependency manipulation. In particular in [10] the problem of determining minimal representations (coverings) of sets of functional dependencies is considered. Here the same problem is stated in much more general terms as the problem of determining minimal representations (coverings) of directed hypergraphs. Various concepts of minimal coverings of directed hypergraphs are introduced and their complexity is discussed. In the case of R-triangular hypergraphs we may wish to determine the minimal coverings with respect to all parameters which may be adopted to characterize a hypergraph (number of hyperarcs, number of adjacency lists required for the representation, length of the overall description,etc.).

In particular, we consider two problems which are the generalization to hypergraphs of the transitive reduction [1] and of the minimum equivalent digraph problem [7] for directed graphs and we show that in the case of hypergraphs both problems are NP-complete while in the case of graphs the transitive reduction is polynomial. Moreover we consider other minimal coverings and we prove their intractability. A problem which is instead shown to be polynomial is the problem of determining a "source minimum" covering, which is shown to be a covering with the minimum number of lists in its representation by means of adjacency lists (and which is equivalent to a problem already considered in the theory of functional dependency in relational data bases [10]). The relationships among the various concepts of minimality are also studied and in particular it is proved that there are coverings which are simultaneously minimal with respect to all criteria.

A formulation based on directed labelled graphs (FD-graphs), previously introduced in [2] is used as a representation of hypergraphs in order to prove the stated results.

In the next paragraph, after providing the basic definitions of R-triangular hypergraphs and their minimal coverings, the main results of the paper concerning the complexity of determining the minimal coverings and their relationships are stated. In Paragraph 3 the formalism of FD-graphs is introduced and the results stated in the preceding paragraph are proved.

Finally, in Paragraph 4 we give examples of applications of hypergraphs and their coverings to functional dependency manipulation [2,10] and to and-or graphs representation [13].

2. HYPERGRAPHS AND THEIR MINIMAL REPRESENTATIONS

Various definitions of hypergraphs have been introduced in the literature (see for example [4,5]). A sufficiently general definition that suits our purpose is the following:

DEFINITION 1. A *generalized hypergraph* is a pair ⟨N,H⟩ where N is the set of *nodes* and H is the set of *hyperarcs*, where a hyperarc is a *structure* which is either a node or a (either ordered or not) set of structures.

Notice that hypergraphs (in the sense of [4]) and directed graphs represent special classes of generalized hypergraphs.

In this paper we will deal with a particular class of generalized hypergraphs where a hyperarc is an ordered pair composed by a set of nodes and a single node.

DEFINITION 2. A generalized hypergraph $\mathcal{H} = \langle N, H \rangle$ is an R-*triangular directed* hypergraph if every hyperarc h $\in$ H is an ordered structure (X,i) where X $\subseteq$ N and i $\in$ N. Given an R-triangular hypergraph we call *source set* a set of nodes that appears as the left side of at least one hyperarc.

From now on we will refer to R-triangular directed hypergraphs simply as hypergraphs.

Example 1. The hypergraph $\mathcal{H} = \langle$ {A,B,C,D,E,F},{ (({A,B},C),({B},D),({C,D},E), ({C,D},F) }$\rangle$ is shown in Fig. 1, where hyperarcs are represented by arrows.

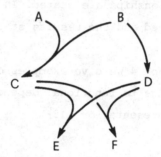

Fig. 1. R-triangular hypergraph.

The basic parameters which will be taken into account in order to evaluate the algorithms presented in this paper will be the following: the number of nodes of the hypergraph (n) the number of hyperarcs (m), the number of source sets (n'), the *source area* (the sum of cardinalities of the source sets,(s) and the overall lenght of description of the hypergraph $|\mathcal{H}|$.

In the previous example we have:

$$n = 6, \quad m = 4, \quad n' = 3, \quad s = 5$$

As far as the lenght of the description is concerned, if we assume a representation based on adjacency lists (where for every source set the

list of adjacent nodes is given) we have $|\mathcal{H}| \underset{=}{\sim}$ s+m. According to the
same representation, the number of source sets, n', corresponds to the
number of adjacency lists.

In order to simplify the notation, here and in the following, nodes
will be denoted by the first latin upper case letters A,B,... and sets
of nodes will be expressed by concatenating the names of nodes (e.g. AB
instead of {A,B} and, in particular, A instead of {A} when no ambiguity
may arise). Besides, the last latin upper case letters X,Y,...,Z will be
used to denote sets of nodes. In this case concatenation will stand for
union (XY stands for X $\cup$ Y) and the cardinality of X will be denoted by
$|X|$.

Since the aim of this paper is to investigate both the concept of
closure and of covering of hypergraphs, the main definition which will
be used throughout the paper concerns paths in a hypergraph.

Several different definitions of path in hypergraphs exist; in our
case we introduce the concept of hyperpath which is derived by extending
the reflexivity and transitivity rules used for the definition of path
in a graph.

DEFINITION 3. Let $\mathcal{H} = \langle N,H \rangle$ be a hypergraph and let $X \subseteq N$, $i \in N$.
There exists a hyperpath $\langle X,i \rangle$ *in* $\mathcal{H}$ *from* X *to* i, *if:*

- (X,i) $\in$ H, or
- i $\in$ X (*extended reflexivity*), or
- there exists a set of nodes Y = $\{n_1,...,n_m\}$ such that there exist hyperpaths $\langle X,n_j \rangle$, for j = 1,...,m in $\mathcal{H}$ and (Y,i) is a hyperarc in H (*extended transitivity*).

Note that when X and Y are singletons, the extended reflexivity and
transitivity rules coincide with the usual definitions of reflexivity
and transitivity as they are defined in graphs. Given the hypergraph $\mathcal{H}$
of Fig. 1 some of the hyperpaths which exist in $\mathcal{H}$ are: $\langle AB,C \rangle$, $\langle AB,A \rangle$,
$\langle AB,E \rangle$.

By means of the previous definition, we may introduce the concept of
closure of a hypergraph.

DEFINITION 4. Given a hypergraph $\mathcal{H} = \langle N,H \rangle$ the *closure* of $\mathcal{H}$, de-noted $\mathcal{H}^+$, is the hypergraph $\langle N,H^+ \rangle$ such that (X,i) is in H^+ iff there exists a hyperpath $\langle X,i \rangle$ in $\mathcal{H}$.

Similarly to what happens in the case of graphs, a problem which arises for the manipulation of hypergraphs is to find a minimal repre-sentation of a hypergraph by means of an other hypergraph which has the same closure but fewer hyperarcs or some other kind of minimality pro-perty. Notice that the problem of finding a minimal representation in the case of hypergraphs is generally more complex than in the case of graphs because, while in the case of graphs the number of arcs in the closure is at most quadratic in the number of nodes, in the case of hypergraphs the number of hyperarcs in the closure is always exponential in the size of N.

DEFINITION 5. Let $\mathcal{H} = \langle N,H \rangle$ be a hypergraph, a *covering* of $\mathcal{H}$ is a hypergraph $\hat{\mathcal{H}} = \langle N,\hat{H} \rangle$ such that $\mathcal{H}^+ = \hat{\mathcal{H}}^+$.

Several concepts of minimal coverings of a hypergraph may be in-troduced.

DEFINITION 6. Given a hypergraph $\mathcal{H} = \langle N,H \rangle$, a *hyperarc* $(X,i) \in H$ is *redundant* if there exists a hyperpath $\langle X,i \rangle$ in $\mathcal{H}' = \langle N,H-\{(X,i)\} \rangle$.

DEFINITION 7. Given a hypergraph $\mathcal{H} = \langle N,H \rangle$ and a hyperarc $(X,i) \in H$, a *node* $j \in X$ is *redundant in* (X,i) if there exists a hyperpath $\langle X-\{j\},i \rangle$ in $\mathcal{H}$.

While, from the definition, the redundancy of a node seems to be relative to a hyperarc, the next result shows under which conditions it is indeed a property of a node itself.

PROPOSITION 1. Let $\mathcal{H} = \langle N,H \rangle$ be a hypergraph and let the node j be redundant in the hyperarc (X,i). If the hyperarc (X,i) is not redundant then the node j is redundant in all hyperarcs (Y,k) such that $Y \supseteq X$.

PROOF. Given all hyperpaths $\langle X-\{j\},i \rangle$ whose existence may be used to show that j is a redundant node, either there exists at least one hyperpath whose existence is based using the hyperarc (X,i) or not. In the second case the hyperarc (X,i) would be redundant. In the first case, by Definition 3, there must exist the hyperpaths ⟨ X-{j},r ⟩ for all r ∈ X. Then, given any hyperarc (Y,k) where Y ⊇ X, we can prove that j is redundant with respect to this hyperarc by showing that there exists a hyperpath ⟨ Y-{j},k ⟩. In fact the hyperpaths ⟨ Y-{j},h ⟩ for every h ∈ X-{j} exist by reflexivity, the hyperpath ⟨ X-{j},j ⟩ exists by hypothesis and hence ⟨ Y-{j},j ⟩ and ⟨ Y-{j},k ⟩ exist by transitivity. This concludes the proof.

                                                              Q.E.D.

DEFINITION 8. A hypergraph $\mathcal{H}$ = ⟨ N,H ⟩ is *nonredundant* if it contains neither redundant hyperarcs nor redundant nodes in the hyperarcs.

Given a hypergraph $\mathcal{H}$ a nonredundant subhypergraph of $\mathcal{H}$ which has the same closure of $\mathcal{H}$ may be obtained by iteratively deleting the redundant arcs and, successively, the redundant nodes until no more redundant arcs and redundant nodes appear. An algorithm for determining a nonredundant subhypergraph of a given hypergraph and its analysis will be given in the next paragraph after introducing a graph formalism for hypergraph representation. The following theorem will then be proved:

THEOREM 1. Given a hypergraph $\mathcal{H}$ = ⟨ N,H ⟩ the problem of determining a nonredundant subhypergraph of $\mathcal{H}$ which has the same closure as $\mathcal{H}$ may be solved in time quadratic in $|\mathcal{H}|$.

Example 2. Given the hypergraph $\mathcal{H}$ of Fig. 2a, a non redundant covering of $\mathcal{H}$, $\mathcal{H}'$ may be obtained by eliminating the redundant hyperarc (A,D) and the redundant node E in the hyperarc (AE,B) (see Fig. 2b).

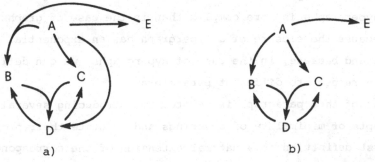

a)   b)

Fig. 2. A nonredundant subhypergraph of a given hypergraph.

Notice that, by changing the order in which redundant arcs and nodes are eliminated, different nonredundant subhypergraphs may be obtained and, in particular, a non redundant subhypergraph with the smaller number of hyperarcs may be derived.

Example 3. In Fig. 3 a nonredundant subhypergraph $\mathcal{H}''$ of the hypergraph given in Fig. 2a is given, which has fewer hyperarcs than the subhypergraph of Fig. 2b.

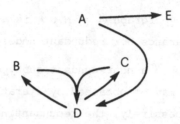

Fig. 3. A nonredundant subhypergraph with the minimum number of
hyperarcs.

Finally we observe that a nonredundant covering with the minimum number of hyperarcs might not be a subhypergraph of the given hypergraph and hence it has to be obtained in a different way.

Example 4. Let us consider the hypergraph $\mathcal{H}'$ of Fig. 2b which is nonredundant. The hypergraph $\mathcal{H}''$ of Fig. 3 is a covering of $\mathcal{H}'$ which is not a subhypergraph of $\mathcal{H}'$ and has the minimum number of hyperarcs.

As we have already observed the problem of determining minimal coverings of hypergraphs is more complex than in the case of graphs essentially because the closure of a hypergraph has an exponential number of hyperarcs and because, in the case of hypergraphs, we can define minimality with respect to different parameters.

The rest of this paragraph is devoted to introducing several different concepts of minimality of coverings and to state their properties.

The first definition is a natural extension of the corresponding definition for graphs.

DEFINITION 9. A *minimum equivalent subhypergraph* of a hypergraph $\mathcal{H}$ is a nonredundant subhypergraph of $\mathcal{H}$ which has the same closure as $\mathcal{H}$ and the minimum number of hyperarcs.

For example the hypergraph $\mathcal{H}$ of Fig. 3 is a minimum equivalent subhypergraph of the hypergraph $\mathcal{H}$ of Fig. 2a. The problem of determining a minimum equivalent subhypergraph of a given hypergraph is NP-complete[*] because this problem (which is clearly in NP) coincides with the problem of the minimum equivalent graph (known to be NP-complete [7]) when for every hyperarc (X,i) we have $|X| = 1$.

In the above definition it is required that the minimal covering which is considered is a subhypergraph of the given hypergraph. In the case of graphs if we drop this condition we obtain a simpler problem (transitive reduction [1]). Analogously, in the case of hypergraphs we may consider minimal coverings which are not required to be subhypergraphs. The first kind of covering, which may be considered the more natural extension to hypergraphs of the transitive reduction is provided in the following definition:

DEFINITION 10. A non redundant covering of a hypergraph $\mathcal{H}$ is said to be a *hyperarc minimum covering* (HM-*covering*) if the number of its hyperarcs is minimum (see again Example 4).

From the computational point of view it is interesting to observe that the complexity of this problem increases dramatically when we go from graphs to hypergraphs. In fact, given a graph $G = \langle N, A \rangle$, the problem of finding the transitive reduction may be solved in polynomial time $O(|N| \cdot |A|)$. Instead, in the case of hypergraphs we will prove the following theorem.

THEOREM 2. Given a hypergraph $\mathcal{H}$, the problem of determining a HM-covering is NP-complete.

(*) Throughout all the paper we will refer to NP-optimization problem and NP-complete optimization problem as defined in [14].

This result which shows that, in the case of hypergraphs, finding the transitive reduction is not simpler than to find the minimum equivalent subhypergraph, suggests taking into consideration other concepts of minimality. In particular, since in the case of hypergraphs coverings may have a different number of source sets with respect to the original hypergraph, the following definition of minimality may be introduced.

DEFINITION 11. A non redundant covering of a hypergraph $\mathcal{H}$ is said to be a *source-minimum covering* (SM-*covering*) if the number of its distinct source sets is minimum.

Due to the reflexivity rule, the number of source sets in the closure of a hypergraph increases exponentially in the number of nodes. It is hence very important to find coverings where such a number is strongly reduced.

Example 5.

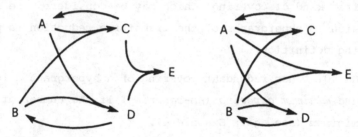

Fig. 4. An SM-covering of a given hypergraph.

In the hypergraph of Fig. 4a the source sets are AB,CD,C,D while in its non redundant covering if Fig. 4b the source sets are AB,C,D and it may be easily seen that there is no covering with fewer source sets.

In the next paragraph the following result will be proved concerning the problem of determining an SM-covering of a given hypergraph:

THEOREM 3. Given a hypergraph $\mathcal{H} = \langle N, H \rangle$, the problem of determining an SM-covering of $\mathcal{H}$ is polynomially solvable and requires time quadratic in $|\mathcal{H}|$.

Note that the definition of SM-covering is not meaningful in the case of graphs, since the fact that the outdegree of a node is zero or not is invariant in all coverings.

By composing the previous definitions and by taking into consideration the source area of a hypergraph as a new parameter to be minimized, the following three concepts of minimality may also be introduced:

DEFINITION 12. A nonredundant covering of a hypergraph is said to be

- a *source-hyperarc-minimum covering* (SHM-*covering*) if it is a HM-covering with the minimum number of source sets;
- an *optimum source-minimum covering* (OSM-*covering*) if it is an SM-covering with the minimum source area;
- an *optimum covering* (O-*covering*) if it is an SHM-covering with the minimum source area.

Example 6. In Figure 5 we have: a) a nonredundant hypergraph $\mathcal{H}$, b) an SM-covering of $\mathcal{H}$ obtained by replacing the hyperarc (CD,E) by the hyperarc (AB,E), (note that such a covering is not hyperarc minimum), c) a HM-covering of $\mathcal{H}$ obtained from it by replacing the hyperarcs (F,E),(E,G), (E,H) by the hyperarcs (F,G),(F,H), (note that such a covering is not source minimum), d) an SHM-covering of $\mathcal{H}$ obtained by combining the above replacements (note that such a covering is not an OSM-covering), e) an OSM-covering of $\mathcal{H}$ obtained from the SM-covering in b) by replacing the hyperarc (HGK,L) by (FK,L), (note that such a covering is neither SHM nor HM), f) an O-covering of $\mathcal{H}$ obtained from the SHM-covering in d) by replacing the hyperarc (HGK,L) by (FK,L). Notice that all such coverings are not subhypergraphs of $\mathcal{H}$.

Given a hypergraph $\mathcal{H}$, finding an SHM-covering and an O-covering of $\mathcal{H}$ are again NP-complete problems. This derives from the fact that in both cases such coverings are also required to be hyperarc minimum coverings by definition and we know that finding a HM-covering is NP-complete. Finding an OSM-covering is also an NP-complete problem as will be shown in the next paragraph:

THEOREM 4. Given a hypergraph $\mathcal{H}$, the problem of determining an

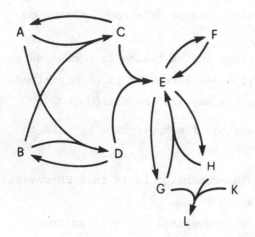

a) a nonredundant hypergraph $\mathcal{H}$

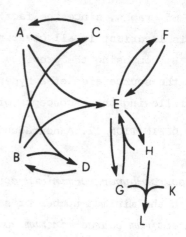

b) a SM-covering of $\mathcal{H}$

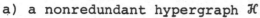

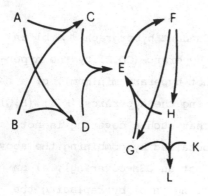

c) a HM-covering of $\mathcal{H}$

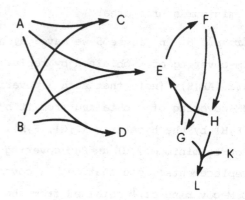

d) a SHM-covering of $\mathcal{H}$

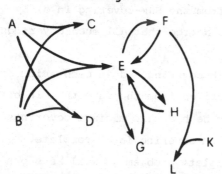

e) an OSM-covering of $\mathcal{H}$

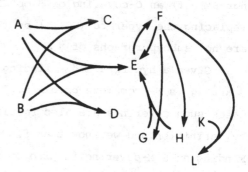

f) an O-covering of $\mathcal{H}$

Fig. 5. Minimal coverings of a hypergraph.

OSM-covering is NP-complete.

On the other hand it is interesting to observe the adequacy of the given concepts of minimality by stating the following theorems which establish the relationships between the various minimality criteria and which will also be proved in the next paragraph.

THEOREM 5. An SHM-covering of a hypergraph $\mathcal{H}$ is also an SM-covering of $\mathcal{H}$.

THEOREM 6. An O-covering of a hypergraph $\mathcal{H}$ is also an OSM-covering of $\mathcal{H}$.

THEOREM 7. A non redundant covering of a hypergraph $\mathcal{H}$ is an OSM-covering if and only if it has the minimum source area among all coverings of $\mathcal{H}$.

The relationships among different types of minimal coverings which are stated in the previous theorems are summarized in the following Figure 6 (where A ⇒ B means that A-minimality implies B-minimality).

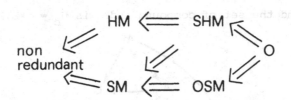

Fig. 6. Relationships among minimal coverings.

Such results imply that the conditions given in Definition 12 (such as the fact that "an SHM-covering is a HM-covering" etc.) are not restrictive and that there exist coverings which are simultaneously minimal with respect to all criteria: number of hyperarcs, number of source sets, source area. In particular it must be noted that the O-covering corresponds to a "minimum lenght" representation of a hypergraph among all possible coverings, if we assume as lenght of a representation the sum of the number of hyperarcs and the source area (as it was suggested at the beginning of this paragraph).

3. GRAPH ALGORITHMS FOR THE MINIMAL REPRESENTATION OF HYPERGRAPHS

Let us now provide the proofs of the results stated in the preceding paragraph. Some of these proofs are based on a graph representation of hypergraphs which has been previously introduced in [2] for the manipulation of functional dependency in relational data bases.

DEFINITION 13. Given a hypergraph $\mathcal{H} = \langle N,H \rangle$ the FD-*graph of* $\mathcal{H}$ is the labelled graph $G_H = \langle N_H, A_f, A_d \rangle$ where:

- $N_H = N \cup N_c$ is a set of *nodes*, where N will be called the set of *simple nodes* and $N_c = \{x \mid \exists \, i \in N$ such that $(X,i) \in H$ and $|x| \neq 1\}$ will be called the set of *compound nodes*;

- $A_f = \{(X,i) \mid$ for every $(X,i) \in H\} \subseteq N_H \times N$ is a set of arcs (labelled f) that will be called the set of *full arcs*;

- $A_d = \{(X,j) \mid$ for every $X \in N_c$ and $j \in X\} \subseteq N_c \times N$ is a set of arcs (labelled d) that will be called the set of *dotted arcs*.

Example 7. Let us consider the hypergraph of Fig. 2a. Its FD-graph representation is given in Fig. 7. In this case the set of simple nodes is $N = \{A,B,C,D,E,\}$ and the set of compound nodes is $N_c = \{AE,BC\}$

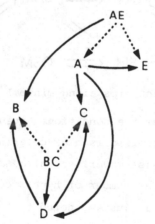

Fig. 7. The FD-graph of the hypergraph in Fig. 2a.

Given a hypergraph $\mathcal{H}$ with n nodes, m hyperarcs, n' source sets, n" source singletons (source sets with cardinality 1) and source area s, it will be represented by a FD-graph with n simple nodes, $n_1 = n'-n''$ compound

nodes, m full arcs and m_1 = s-n" dotted arcs. If we consider the lenght
of the description of the FD-graph we may easily assume that it coincides
with the lenght $|\mathcal{H}| \simeq$ s+m of the description of the given hypergraph.

The use of FD-graphs and of their closure in some cases allows to
determine minimal coverings of hypergraphs without falling into the exponential explosion of the hypergraph closure because the FD-graph closure grows only at most quadratically. More precisely in order to find
a covering of a hypergraph $\mathcal{H}$ with suitable minimality properties we first
give the FD-graph representation G_H of the given hypergraph, then we determine the closure of G_H (instead of $\mathcal{H}^+$) in order to provide the minimal
covering $\mathcal{H}'$.

The sequence of transformations that we may go through in such cases is
given in Fig. 8 (continuous line).

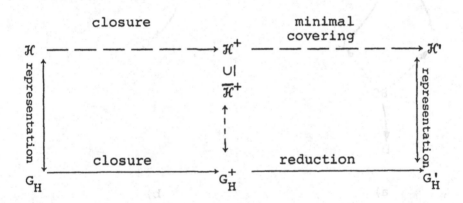

Fig. 8. The sequence of transformations to determine minimal coverings.

First of all let us define the concept of FD-path that will be used in
order to define the closure of an FD-graph.

DEFINITION 14. Given an FD-graph $G_H = \langle N_H, A_f, A_d \rangle$ and a pair of
nodes $i,j \in N_H$, an FD-*path* $\langle i,j \rangle$ from i to j is a minimal subgraph
$\bar{G}_H = \langle \bar{N}_H, \bar{A}_f, \bar{A}_d \rangle$ of G_H such that $i,j \in \bar{N}_H$ and either $\bar{A}_f \cup \bar{A}_d = \{i,j\}\}$ or
one of the following possibilities holds:

- j is a simple node and there exists a node k such that $(k,j) \in \bar{A}_f \cup \bar{A}_d$

and there is an FD-path $\langle i,k \rangle$ in $\bar{G}_H$ (*transitivity*);

- j is a compound node with component nodes $m_1,\ldots,m_r$ and $(j,m_1),\ldots,$ $(j,m_r) \in \bar{A}_d$ and there are FD-paths $\langle i,m_1 \rangle,\ldots,\langle i,m_r \rangle$ in $\bar{G}_H$ (*union*).

Furthermore an FD-path $\langle i,j \rangle$ is *dotted* if all its arcs leaving i are dotted, otherwise it is *full*.

Example 8. In Fig. 9a a full FD-path and in Fig. 9b a dotted FD-path from the hypergraph of Fig. 7 are shown:

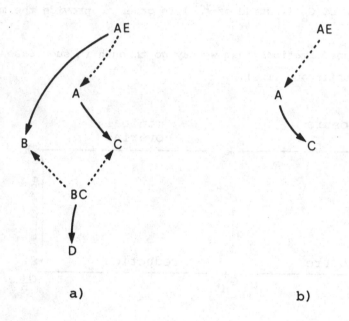

a)   b)

Fig. 9. Examples of FD-paths.

DEFINITION 15. Given an FD-graph $G_H = \langle N_H, A_f, A_d \rangle$ we define *closure of* G_H the labelled graph $G_H^+ = \langle N_H, A_f^+, A_d^+ \rangle$ where an arc (i,j) is

a) in A_d^+ iff there exists a dotted FD-path $\langle i,j \rangle$

b) in A_f^+ iff (i,j) $\notin A_d^+$ and there exists a full FD-path $\langle i,j \rangle$.

In [2] an algorithm is shown which, given an FD-graph G_H and a node i, provides the sets of nodes which may be reached from the node i by means of a full or dotted FD-path (in time $O(m_H)$, where $m_H = m+m_1$ is the number of arcs of G_H and $m(m_2)$ is the number of full (dotted) arcs.

Such algorithm is an extension to FD-graphs of the usual transitive closure algorithm for graphs. The only substantial modification concerns the application of the union rule (see Definition 14) that is implemented by associating a counter with every compound node j. This counter keeps track of the number of component nodes of j which are currently reached from the source node i. By applying this algorithm to all the nodes, we may determine the closure of G_H in time $O(\bar{n}_H \cdot m_H)$ where $\bar{n}_H$ is the total number of nodes of G_H with at least one outgoing arc. In terms of the parameters of the hypergraph we have that the closure algorithm runs in time $O(n' \cdot |\mathcal{H}|)$ since $m_H = m+m_1 \leq m+s \cong |\mathcal{H}|$ and $\bar{n}_H = n'$.

The closure of an FD-graph is a succint representation of the closure of the corresponding hypergraph in the sense expressed in the following theorem:

THEOREM 8. Let $\mathcal{H} = \langle N,H \rangle$ be a hypergraph and $G_H = \langle N_H, A_f, A_d \rangle$ the corresponding FD-graph. Given a pair of nodes $i,j \in N_H$ where j is a simple node, the arc (i,j) is in G_H^+ if and only if there exists a corresponding hyperarc in $\mathcal{H}^+$.

PROOF. *Only if part*: since every arc in G_H^+ incident into a simple node is either in G_H or is derived by applying the transitivity and the union rules, it is easy to observe that the corresponding hyperarc is either in $\mathcal{H}$ or is derived in $\mathcal{H}^+$ by applying the extended transitivity and the reflexivity rules.

If part: if a hyperarc (i,j) is in $\mathcal{H}^+$, where j is a single node, the hyperpath $\langle i,j \rangle$ is in $\mathcal{H}$. By induction on the structure of a hyperpath the following cases may arise (the first two cases are the basis of the induction):

- either (i,j) is a hyperarc of $\mathcal{H}$, then $\langle i,j \rangle$ is an FD-path in G_H;

- or (i,j) is a hyperarc of $\mathcal{H}^+$ obtained by reflexivity, then the dotted arc (i,j) appears in G_H;

- or there exists a set of simple nodes $Y = \{n_1, \ldots, n_m\}$ such that $\langle i, n_k \rangle$
  for $k = 1, \ldots, m$ are hyperpaths and (Y, j) is a hyperarc. In this case,
  by inductive hypothesis there exist FD-paths $\langle i, n_k \rangle$ in G_H for
  $k = 1, \ldots, m$; then by union rule (or by transitivity rule if $m = 1$) there
  exists the FD-path $\langle i, j \rangle$.

Q.E.D.

Notice that the subhypergraph of $\mathcal{H}^+$ whose hyperarcs exist if and
only if the corresponding arc exists in G_H^+ is the hypergraph that we
have denoted $\overline{\mathcal{H}}^+$ in figure 8 and is itself a covering of $\mathcal{H}$.

The next step toward the determination of minimal coverings of a
hypergraph will make use of the FD-graph representation of a hypergaph.
Starting from the closure of the FD-graph we will apply transformation
rules that bring the FD-graph into reduced forms and we will show that
such reduced forms correspond to the FD-graphs of a nonredundant covering
and of an SM-covering of the original hypergraph (see again Fig. 8).

Let us first introduce the following definition.

DEFINITION 16. Given an FD-graph we say that i) a *compound node* k
is *redundant* if for every full arc (k,j) there exists a dotted FD-path
$\langle k, j \rangle$; ii) a *dotted (full) arc* (k,j) is *redundant* if there exists a
dotted (full or dotted) FD-path $\langle k, j \rangle$ which does not contain the arc
(k,j).

By means of the following proposition, which is a straightforward
consequence of Theorem 8, we give the first rules for reducing an FD-graph
(see Fig. 8). Such rules allow us to find a nonredundant subhypergraph of
a given hypergraph.

PROPOSITION 2. Given an FD-graph G_H, every FD-graph obtained from
G_H by eliminating any redundant node together with all its outgoing arcs
or any redundant arc, is the representation of a covering of the hypergraph represented by G_H.

From now on, by "elimination of a redundant node in an FD-graph" we
will mean also the elimination of all arcs leaving the redundant node
(notice that, by definition of FD-graph, compound nodes do not have in-

coming arcs).

Now we prove that, as we stated in the preceding paragraph, a non-redundant subhypergraph of a given hypergraph may be found in time $O(|\mathcal{H}|^2)$:

PROOF OF THEOREM 1. Given a hypergraph $\mathcal{H}$ we

1. determine the FD-graph G_H corresponding to $\mathcal{H}$;

2. eliminate redundant nodes from G_H by determining the closure of G_H;

3. eliminate redundant full arcs from G_H;

4. eliminate redundant dotted arcs from G_H;

5. derive the hypergraph $\mathcal{H}'$ corresponding to the reduced FD-graph.

In order to show the correctness of such algorithm we first observe that $\mathcal{H}'$ is a covering of $\mathcal{H}$ (by Proposition 2); then we show that $\mathcal{H}'$ has no redundancies. In fact, if there was a redundant hyperarc (X,i) in the hypergraph $\mathcal{H}'$, there would be either a redundant full arc (X,i) in the FD-graph or at least a redundant compound node, namely the node X itself (contradiction). On the other side, if there was a redundant node $j \in X$ in the hypergraph, with respect to some nonredundant hyperarc (X,i) then there would be a redundant dotted arc (X,j) in the FD-graph. This means that the elimination of redundancies in the FD-graph implies the elimination of redundancies in $\mathcal{H}$. Concerning the efficiency of the given procedure, since both steps 1 and 5 require linear time in the size of the input, the cost is essentially due to steps 2,3 and 4. Step 2 requires time $O(n' \cdot |\mathcal{H}|)$, where n' is the number of source sets and $|\mathcal{H}|$ is the length of the description of $\mathcal{H}$, because the elimination of redundant nodes in the FD-graph is immediately deduced from the closure (in fact a compound node is redundant if and only if all its outgoing arcs in the closure are dotted). The elimination of redundant full arcs requires that for every full arc (h,i) we determine in time $O(|\mathcal{H}|)$ all full FD-paths starting from the node h and which do not include the arc (h,i). The overall cost of step 3 is hence $O(m \cdot |\mathcal{H}|)$.

Finally, in order to eliminate redundant dotted arcs we may proceed in the following way: given any dotted arc (h,i) we may compute the set

of nodes j such that a dotted H-path $\langle h,j \rangle$ exists which does not include
the dotted arc (h,i). If i belongs to such a set this means that the arc
(h,i) is redundant. Since the time required to answer this question is
$O(|\mathcal{H}|)$ and since the number of dotted arcs is $m_1 = s{-}n'' \leq s$, the overall
cost of step 4 is $O(s \cdot |\mathcal{H}|)$. Taking into account that $n' \leq s$ and that
$|\mathcal{H}| = s{+}m$ we obtain that the cost of determining a nonredundant sub-
hypergraph is $O(|\mathcal{H}|^2)$.

Q.E.D.

A second rule for the reduction of FD-graphs will now be introduced.
Let us first consider the following definition.

DEFINITION 17. Given an FD-graph $G_H = \langle N_H, A_f, A_d \rangle$:

- a pair of nodes $i,j \in N_H$ are said to be *equivalent* if both the (full
  or dotted) arcs (i,j) and (j,i) belong to the closure of G_H;
- a compound node i is said to be *superfluous* if there exists a dotted
  FD-path $\langle i,j \rangle$ where j is equivalent to i.
- G_H is *LR-minimum*[(*)] if it has neither redundant nodes and arcs nor
  superfluous nodes.

Example 9. In the FD-graph of Figure 10a) (corresponding to the hypergraph of Figure 4a), the nodes AB and CD are equivalent and the node CD
is superfluous. The FD-graph in Fig. 10b) is LR-minimum.

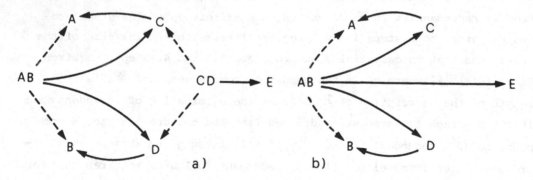

Fig. 10. FD-graph containing a superfluous node and LR-minimum FD-graph.

(*) The name is due to the properties of a corresponding definition given
      in [10] for functional dependencies.

The rule of elimination of a superfluous node i from an FD-graph consists in eliminating i together with all its outgoing dotted arcs and moving, at the same time, all its outgoing full arcs to an equivalent node j which is connected to i by a dotted FD-path.

PROPOSITION 3. Let $G_H = \langle N_H, A_f, A_d \rangle$ be the FD-graph representation of a hypergraph $\mathcal{H}$. Let i be a superfluous node in G_H and let j be a node equivalent to i such that there exists a dotted FD-path $\langle i, j \rangle$. Let $G_{H'} = \langle N_H', A_f', A_d' \rangle$ be an FD-graph where:

- $A_f' = (A_f \cup \{(j,k) \,|\, (i,k) \in A_f \text{ and } k \in N_H\}) \setminus \{(i,k) \,|\, k \in N_H\}$

- $A_d' = A_d \setminus \{(i,k) \,|\, k \in N_H\}$

- $N_H' = N_H \setminus \{i\}$.

Then $G_{H'}$ is the FD-graph representation of a covering of the hypergraph $\mathcal{H}$.

PROOF. Starting from G_H, we construct the FD-graph $G_{H''}$ by adding the redundant arc (j,h) for every (i,h) in A_f. Since G_H can be obtained from $G_{H''}$ by eliminating the redundant full arcs introduced above by Proposition 2 $\mathcal{H}''$ is a covering of $\mathcal{H}$. Since the node i is redundant in $G_{H''}$, because for every full arc (i,h) there is a dotted FD-path $\langle i, h \rangle$ passing through j, $G_{H'}$ can be obtained from $G_{H''}$ by eliminating the node i and all its outgoing arcs. By proposition 2, $\mathcal{H}'$ is a covering of $\mathcal{H}''$ and therefore of $\mathcal{H}$ as well.

Q.E.D.

From now on, by "elimination of a superfluous node" we will mean the procedure indicated in Proposition 3 (for instance, the FD-graph in Fig. 10b is obtained from the FD-graph in Fig. 10a by eliminating the superfluous node CD).

Before proceeding in proving the results stated in paragraph 2, we need the following two lemmata. Lemma 1 outlines a structural property of LR-minimum FD-graphs which allows to establish (by Lemma 2) a strong correspondence between LR-minimum FD-graphs and source minimum hypergraphs which will be needed for most of the subsequent results. Furthermore Lemma

1 will also be used to prove Theorem 6. Notice that all structural properties of LR-minimum FD-graphs stated in [2] can be easily derived from
Lemma 1 but the viceversa does not hold.

LEMMA 1. Let $G_{H'} = \langle N_{H'}, A_f', A_d' \rangle$, $G_{H''} = \langle N_{H''}, A_f'', A_d'' \rangle$ be the FD-graph
representations of two coverings of a hypergraph $\mathcal{H}$. If both $G_{H'}$ and $G_{H''}$
are LR-minimum then there exists a bijection $\phi: N_{H''} \to N_{H'}$ such that, for
every node $i \in N_{H''}/N_{H'}$

a) $\phi(i) \in N_{H'}/N_{H''}$

b) $\phi(i)$ is equivalent to i in $G_{H'''} = \langle N_{H'} \cup N_{H''}, A_f', A_d' \cup A_d'' \rangle$

c) there exists a dotted FD-path $\langle \phi(i), i \rangle$ in $G_{H'''}$

PROOF. In order to prove the lemma we need first to prove the following claim.

CLAIM 1. Let $\mathcal{H}_1$ and $\mathcal{H}_2$ be two coverings of $\mathcal{H}$, G_{H_1} and G_{H_2} the
corresponding FD-graphs, $G_{H_1}^+ = \langle N_{H_1}, A_{1f}^+, A_{1d}^+ \rangle$ and $G_{H_2}^+ = \langle N_{H_2}, A_{2f}^+, A_{2d}^+ \rangle$
be their closures and let i,j be in $N_{H_1} \cap N_{H_2}$. If (i,j) is in A_{1f}^+ and
(i,j) is in A_{2d}^+, then every dotted FD-path $\langle i,j \rangle$ in G_{H_2} contains a
node k equivalent to i.

PROOF. Let $G_{H_3}^+ = \langle N_{H_3}, A_{3f}^+, A_{3d}^+ \rangle$ be the closure of $G_{H_3} = \langle N_{H_1} \cup N_{H_2}, A_{1f}, A_{1d} \cup A_{2d} \rangle$. Every FD-path $\langle k, \ell \rangle$ in G_{H_3} such that k, ℓ are in
N_{H_1} is also in G_{H_1} and viceversa because G_{H_3} contains more compound
nodes besides all compound nodes in N_{H_1} but not more full arcs. Hence
(i,j) is also in A_{3f}^+. Let us now consider any dotted FD-path $\langle i,j \rangle$ in
G_{H_2} and let $k_1, \ldots, k_s$ be the intermediate nodes on $\langle i,j \rangle$. We have to
prove that at least one of these intermediate nodes is equivalent to i.
By Theorem 8, if (k_e, k_r) is in $A_{2f} \cup A_{2d}$, there exists an FD-path
$\langle k_e, k_r \rangle$ in G_{H_3} because H_3 is a covering of H_1. Moreover, since no dotted
FD-path $\langle i,j \rangle$ is in G_{H_3}, some FD-path $\langle k_e, k_r \rangle$ contains a full arc

leaving i. By Definition 14, in G_{H_3} there exist FD-paths $\langle i,k_e \rangle$ (because

the FD-path $\langle i,j \rangle$ contains k_e) and $\langle k_e,i \rangle$ (by Theorem 8 because the

FD-path $\langle k_e,k_r \rangle$ in G_{H_3} contains the node i). Hence i and k_e are equi-

valent.

END OF PROOF OF CLAIM 1.

Let us now consider the two LR-minimum FD-graphs $G_{H'}$ and $G_{H''}$ as

defined in the statement of the Lemma. We construct the bijection ϕ in

the following way: when $i \in N_{H''} \cap N_{H'}$, then $\phi(i) = i$. Otherwise, if

$\bar{N}_{H''} = N_{H''}\backslash N_{H'}$ is not empty and $i \in \bar{N}_{H''}$, then $\phi(i) = j$ where $j \in \bar{N}_{H'} =$

$= N_{H'}\backslash N_{H''}$ and is derived in the following way.

Let us construct the FD-graph $G_{\bar{H}'}$, by adding i and its outgoing

dotted arcs to $G_{H'}$. Let $G_{\bar{H}'}^+ = \langle N_{\bar{H}'}, \bar{A}_f'^+, \bar{A}_d'^+ \rangle$ be the closure of $G_{\bar{H}'}$. By

Proposition 2, the hypergraph $\mathcal{H}'$ is a covering of $\mathcal{H}'$ and then of $\mathcal{H}''$.

Since i is non redundant in $G_{H''}$ by hypothesis, there exists at least one

simple node r such that (i,r) is in A_f'' and then in $A_f''^+$. Instead, by con-

struction, the arc (i,r) is in $\bar{A}_d'^+$. Hence, by Claim 1, every dotted

FD-path $\langle i,r \rangle$ in $G_{\bar{H}'}$ contains at least a node k equivalent to i. Let

j be a node in $N_{H'}$ equivalent to i and such that there is a dotted

FD-path $\langle i,j \rangle$ in $G_{\bar{H}'}$ that does not contain any other node equivalent to

i. We may show that j is indeed in $\bar{N}_{H'}$ by contradiction. In fact if j was

in $N_{H'} \cap N_{H''}$, by Claim 1 the arc (i,j) would be in $A_d''^+$ and i would be

superfluous in $G_{H''}$ (contradiction with the hypothesis that $G_{H''}$ is

LR-minimum). We prove that ϕ is bijective by showing that ϕ is injective

and that $|\bar{N}_{H''}| = |\bar{N}_{H'}|$. Let $\hat{i}$ be a node in $\bar{N}_{H''}$ different from i. Let us

suppose, by contradiction, that $\phi(\hat{i}) = \phi(i) = j$. Let $G_{\bar{H}''}$ be the FD-graph

obtained from $G_{H''}$ by adding the node j and all its outgoing dotted arcs

and let $G_{\bar{H}''}^+ = \langle N_{\bar{H}''}, \bar{A}_f''^+, \bar{A}_d''^+ \rangle$ be the closure of $G_{\bar{H}''}$. Since j is equiva-

lent to i and $\hat{i}$ in $G_{\bar{H}''}$, (j,i) and $(j,\hat{i})$ are in $\bar{A}_d''^+$. Furthermore, since

the dotted FD-paths $\langle i,j \rangle$ and $\langle \hat{i},j \rangle$ in $G_{\bar{H}'}$ do not contain other nodes

equivalent to i or $\hat{i}$ and since $\bar{H}'$ is a covering of H", by Claim 1 there

exist also dotted FD-paths $\langle i,j \rangle$ and $\langle \hat{i},j \rangle$ in $G_{\bar{H}''}$. Now, without loss of

generality, we suppose that the dotted FD-path $\langle j,i \rangle$ in $G_{\bar{H}''}$ does not
contain the node $\hat{i}$ (otherwise we could refer to $\langle j,\hat{i} \rangle$). Since in $G_{\bar{H}''}$
there exist the dotted FD-paths $\langle \hat{i},j \rangle$ and $\langle j,i \rangle$ and since $\langle j,i \rangle$ does
not contain the node $\hat{i}$, there exists also a dotted FD-path $\langle \hat{i},i \rangle$ in $G_{\bar{H}''}$.
This FD-path is also in $G_{H''}$ because $G_{\bar{H}''}$ differs from $G_{H''}$ only in the
compound node j that has no outgoing full arcs. Hence $\hat{i}$ is superfluous
in $G_{H''}$ and we get a contradiction with the hypothesis that $G_{H''}$ is
LR-minimum. Therefore the mapping ϕ is injective and, hence, $|\bar{N}_{H''}| \leq$
$\leq |\bar{N}_{H'}|$. If we exchange $G_{H'}$ with $G_{H''}$ and viceversa in our argument, we
obtain also $|\bar{N}_{H'}| \leq |\bar{N}_{H''}|$. Hence forth $|\bar{N}_{H'}| = |\bar{N}_{H''}|$ and ϕ is a bijection.
This concludes the proof of Part a) of the lemma. In order to prove Part
b) we observe that since i and $j = \phi(i)$ are equivalent in $G_{\bar{H}'}$ and since
$\bar{H}'$ is a covering of H''' by Proposition 2, i and j are equivalent in $G_{H'''}$
by Theorem 8. Finally we have to prove Part c), i.e. that there is a
dotted FD-path $\langle j,i \rangle$ in $G_{H'''}$. First of all we notice that in $G_{\bar{H}''}$ there
is a dotted FD-path $\langle j,i \rangle$. We show that this FD-path does not contain
any node equivalent to j by contradiction. Let us suppose that there exists a node $\hat{i}$ equivalent to j in $\langle j,i \rangle$. Without loss of generality we
can suppose also that the dotted FD-path $\langle j,\hat{i} \rangle$ does not contain any node
equivalent to j. Since there is also a dotted FD-path $\langle i,j \rangle$ in $G_{\bar{H}''}$ (as
we have already proved) there exists a dotted FD-path $\langle i,\hat{i} \rangle$ in $G_{\bar{H}''}$.
This FD-path is also in $G_{H''}$ and i is superfluous (contradiction with the
hypothesis that $G_{H''}$ is LR-minimum). Therefore the dotted FD-path $\langle j,i \rangle$
in $G_{\bar{H}''}$ does not contain any node equivalent to j. Hence, since $\bar{H}''$ is a
covering of H''', by Claim 1 there is a dotted FD-path $\langle j,i \rangle$ in $G_{H'''}$ and
this concludes the proof.

                                                                         Q.E.D.

The next lemma establishes the correspondence between LR-minimum
FD-graphs and source-minimum hypergraphs. This result is useful both for
applying to source minimum hypergraphs the computational results proved
in [2] for LR-minimum FD-graphs and for deriving other results stated in
section 2, concerning the other concepts of minimal coverings.

LEMMA 2. A hypergraph $\mathcal{H}$ is source minimum iff its FD-graph representation is LR-minimum.

PROOF. *Only if part*. Let G_H be the FD-graph representation of $\mathcal{H}$. G_H does not contain redundant or superfluous nodes, because otherwise we could reduce the number of source sets in $\mathcal{H}$ by eliminating either the redundant or the superfluous nodes (by Propositions 2 and 3). G_H does not have redundant arcs because otherwise $\mathcal{H}$ would be redundant (by Theorem 8). Hence G_H is LR-minimum.

If part. Let $G_H = \langle N_H, A_f, A_d \rangle$ be the LR-minimum FD-graph representation of the hypergraph $\mathcal{H}$. Let $\mathcal{H}'$ be an SM-covering of $\mathcal{H}$ and let $G_{H'} = \langle N_{H'}, A_f', A_d' \rangle$ be its FD-graph representation. By the only if part of this lemma $G_{H'}$ is LR-minimum. Hence, by lemma 1, G_H and $G_{H'}$ have the same number of nodes with outdegree > 0. This means that $\mathcal{H}$ and $\mathcal{H}'$ have the same number of source sets, i.e. $\mathcal{H}$ is source minimal.

Q.E.D.

We are now able to prove that the problem of determining a source minimum covering of a given hypergraph $\mathcal{H}$ can be solved in time quadratic in the size of the description of $\mathcal{H}$.

PROOF OF THEOREM 3. An SM-covering of a given hypergraph $\mathcal{H}$ may be obtained by the following steps:

1. determine the FD-graph representation of $\mathcal{H}$;

2. eliminate redundant nodes;

3. eliminate superfluous nodes;

4. eliminate redundant arcs;

5. derive the hypergraph $\mathcal{H}'$ corresponding to the reduced FD-graph.

By Proposition 2 and 3, $\mathcal{H}'$ is a covering of $\mathcal{H}$. By Lemma 2, $\mathcal{H}'$ is source minimum. As far as the complexity is concerned it has been shown in the proof of Theorem 1 that steps 1,2,4 and 5 require time quadratic in $|\mathcal{H}|$. Superfluous nodes can be easily recognized from the closure. Hence the whole complexity of the algorithm remains quadratic in $|\mathcal{H}|$. Actually in [2] a more efficient implementation of this algorithm was given, that requires time $O(t \cdot |\mathcal{H}|)$ where t $(n' \leq t \leq |\mathcal{H}|)$ is a parameter

depending on the structure of the hypergraph which takes value n' (number
of nodes) when the hypergraph is indeed a graph.

                                                                    Q.E.D.

After having considered the complexity properties of nonredundant
and source minimum coverings let us now turn to the harder problems.

First of all we provide the NP-completeness proofs for the hyperarc
minimum and source optimum coverings.

PROOF OF THEOREM 2. In order to prove that the problem of determining an HM-covering is NP-complete we may give a polynomial reduction from
the set-covering problem to the problem of minimizing the number of hyperarcs of a nonredundant covering of a hypergraph, analogously to what is
done in [9]. Let an instance of the set covering problem be given: let
$S = \{s_1, \ldots, s_n\}$ be a set of elements and $S_1, \ldots, S_m$ be a family of subsets
of S such that $\bigcup_{i=1}^{m} S_i = S$. The set-covering problem is the problem of
finding a subfamily minimizing the number of sets $S_{i_1}, \ldots, S_{i_k}$ such that
$\bigcup_{j=1}^{k} S_{i_j} = S$. Given the above instance we may construct a hypergraph whose
nodes are $\bar{s}_1, \ldots, \bar{s}_n, \bar{S}_1, \ldots, \bar{S}_m, \bar{T}$ and for every $s_j \in S_i$ there is a cor-
responding hyperarc $(\bar{S}_i, \bar{s}_j)$; besides there are the hyperarcs $(\{\bar{s}_1, \ldots, \bar{s}_n\}, \bar{S}_i)$ and the hyperarcs $(\bar{T}, \bar{S}_i)$ for all $i = 1, \ldots, m$ (see Fig. 11).

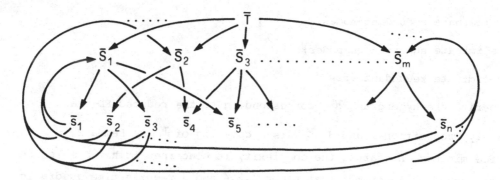

Fig. 11. Hypergraph associated with an instance of the set-covering pro-
         blem.

Note that if from the hypergraph the node $\bar{T}$ and the hyperarcs leaving it are taken out, the remaining hypergraph is nonredundant and no

covering with a smaller number of arcs may exist. Hence the only redundant arcs may be those leaving $\bar{T}$ (*). In fact if the sets $S_{i_1}, \ldots, S_{i_k}$ provide a covering of S, all arcs leaving $\bar{T}$ and different from $(\bar{T}, \bar{S}_{i_1}) \ldots, (\bar{T}, \bar{S}_{i_k})$ are redundant and may be eliminated without changing the closure. Since the reduction from the instance of set covering to the instance of HM-covering of the hypergraph is polynomial we have shown that if we know how to minimize the number of hyperarcs in the hypergraph we would solve the set covering problem. Hence the hyperarcs minimization problem is NP-hard. The easy observation that such problem is solvable in polynomial nondeterministic time completes the proof.

Q.E.D.

PROOF OF THEOREM 4. In order to prove that the problem of determining an OSM-covering is NP-complete we may use a slight modification of the proof of Theorem 2. Let us again consider the hypergraph in Figure 11.

First we add a new node $\bar{T}_1$ and then we replace the hyperarcs $(\bar{T}, \bar{S}_1), \ldots, (\bar{T}, \bar{S}_m)$ with the hyperarc $(\bar{S}_1, \ldots, \bar{S}_m \bar{T}_1, T)$. This latter hyperarc may contain redundant nodes. If we eliminate such nodes by Lemma 2 we obtain a SM-covering of $\mathcal{H}$ since its FD-graph representation G_H is LR-minimum. In fact in G_H neither nodes nor arcs are redundant and no node is superfluous because there are no equivalent node. Notice that if we did not add the node $\bar{T}_1$ in G_H, the node $\bar{S}_1 \ldots \bar{S}_m$ would have been superfluous with respect to the node $\bar{s}_1 \ldots \bar{s}_n$. The OSM-covering of this hypergraph is an SM-covering from which we have eliminated the maximal number of redundant nodes in the previous hyperarc. Hence by determining the OSM-covering we would also solve the set covering problem.

Q.E.D.

Finally we prove the results concerning implications among minimality concepts (see Fig. 6).

(*) Note that in this case the HM-covering problem coincides with the minimum equivalent subhypergraph problem and the relaxation of the subhypergraph constraint does not make the problem simpler. Actually the same theorem can also be used to prove the NP-completeness of the minumum equivalent subhypergraph problem.

Let us first of all prove that since source minimality is implied by source-hyperarc minimality an SHM-covering may be found among SM-coverings.

PROOF OF THEOREM 5. Let a hypergraph $\mathcal{H}$ be given. Let $\mathcal{H}'$ be an SHM-covering of $\mathcal{H}$; in order to prove that it is also an SM-covering let us proceed by contradiction. Let us suppose that $\mathcal{H}'$ is not an SM-covering of $\mathcal{H}$ and let $G_{H'}$ be the FD-graph associated to $\mathcal{H}'$. By Lemma 2, $G_{H'}$ is not LR-minimum. Since H' is nonredundant, $G_{H'}$ has neither redundant nodes nor redundant arcs. Hence $G_{H'}$ has at least one superfluous node. By eliminating such a node we would determine an FD-graph $G_{H''}$ which represents a hypergraph $\mathcal{H}''$ which is a H-minimum covering of $\mathcal{H}$ but with a smaller number of source sets (contradiction).

Q.E.D.

The second result concerns the fact that an optimal covering may be found among optimal source minimum coverings.

PROOF OF THEOREM 6. Let a hypergraph $\mathcal{H}$ be given. Let $\mathcal{H}'$ and $\mathcal{H}''$ be respectively an O-covering and an OSM-covering of $\mathcal{H}$. By definition of O-covering, in order to prove the theorem, it is sufficient to find an SHM-covering $\mathcal{H}'''$ which has the same source area as $\mathcal{H}''$. To this goal we consider the FD-graphs $G_{H'} = \langle N_{H'}, A_f', A_d' \rangle$ and $G_{H''} = \langle N_{H''}, A_f'', A_d'' \rangle$ associated to $\mathcal{H}'$ and $\mathcal{H}''$. We construct the FD-graph $G_{H'''} = \langle N_{H'} \cup N_{H''}, A_f', A_d' \cup A_d'' \rangle$. By Proposition 2, $\mathcal{H}'''$ is covering of $\mathcal{H}$. By Lemma 1 every compound node in $N_{H'} \setminus N_{H''}$ is superfluous in $G_{H'''}$. By eliminating such nodes we obtain an FD-graph with the same number of full arcs as $G_{H'}$ and the same set of nodes as $G_{H''}$. Therefore the hypergraph represented by this FD-graph is an SHM-covering of $\mathcal{H}$ which has the same source area as $\mathcal{H}''$.

Q.E.D.

Finally we prove that a nonredundant covering with the minimum source area is necessarily a source-minimum covering.

PROOF OF THEOREM 7. Let $\mathcal{H}$ be a hypergraph and let $\mathcal{H}'$ be a covering of $\mathcal{H}$ with the smallest source area. We may show that $\mathcal{H}'$ is also source-minimum. Without loss of generality we assume that $\mathcal{H}'$ is nonredundant (in fact, if $\mathcal{H}'$ was redundant, by eliminating redundancies we could obtain a nonredundant covering of $\mathcal{H}$ with the same source area as $\mathcal{H}'$). Let $G_{H'}$ be the FD-graph representation of $\mathcal{H}'$. Since $\mathcal{H}'$ is nonredundant, $G_{H'}$ has neither redundant nodes nor redundant arcs. Moreover $G_{H'}$ has no superfluous nodes because otherwise, by eliminating such nodes, we would find the FD-graph representation of a covering of $\mathcal{H}'$ with a smaller source area. Hence $G_{H'}$ is LR-minimum and by Lemma 2, $\mathcal{H}'$ is source-minimum.

                                                                    Q.E.D.

4. APPLICATIONS OF MINIMAL REPRESENTATIONS OF HYPERGRAPHS

FD-graphs were first introduced in [2] in connection with the representation and manipulation of sets of *functional dependencies* in relational data bases [15].

In this case, as it was shown in the mentioned reference, the problem is to determine a minimal covering of a set of functional dependencies $X_i \rightarrow Y_i$, $1 \leq i \leq k$, defined over a set of attribute names U, where X_i and Y_i are subsets of U.

For example if A,B,C,D,E are attribute names, the following set of functional dependencies:

$$AB \rightarrow CD \,, \quad B \rightarrow E \,, \quad E \rightarrow C$$

represents the implication between attribute values, that is the pair of values over A and B univokely determine the values over C and D, etc.

Given a set F of functional dependencies, a set of inference rules allows to determine the set F^+ of all dependencies which may be derived as consequences of F. A central problem in relational theory is hence to determine a covering F' of F such that $F'^+ = F^+$ and F' is "minimal" with respect to some criteria [10]. By associating nodes to attributes and hyperarcs $(A_1 \ldots A_n, B_1), \ldots, (A_1 \ldots A_n, B_m)$ to every functional dependency $A_1 \ldots A_n \rightarrow B_1 \ldots B_m$, we may represent a set of functional dependencies by

an R-triangular hypergraph. In [2] it has been shown that the problem of
determining minimal coverings such as the ones considered in [10] may be
efficiently solved by using FD-graphs and their manipulation algorithms.
In particular such minimal coverings correspond to our source-minimum
hypergraphs.

Other kinds of dependencies in relational data base theory such as
the *existence constraints* introduced in [11] may also be treated by using
hypergraph algorithms since inference rules for this kind of constraints
have the same structure as inference rules for functional dependencies.

FD-graphs may also be applied for the efficient manipulation of
AND-OR graphs [13]. In fact it is easy to see that these structures,
which are used for the representation of reduction of problems in problem
solving, may still be represented by hypergraphs (actually L-triangular
hypergraphs since every hyperarc is usually directed from one node, problem to be solved, to a set of nodes, subproblems whose solution is required in order to solve the given problem).

Also in this case the problem of determining minimal descriptions
of AND-OR graphs may arise.

In order to represent and manipulate an AND-OR graph by means of a
FD-graph we may consider the reversed (R-triangular) hypergraph which is
obtained by reversing all hyperarcs.

In Figure 12 we show an AND-OR graph and its representation by means
of an FD-graph

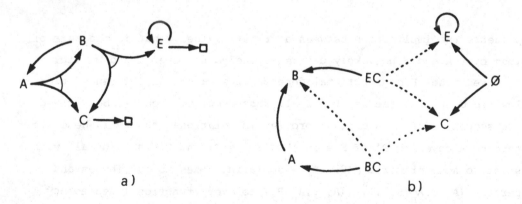

Fig. 12. AND-OR graph and its FD-graph representation.

In this case an FD-path from the empty compound node to a target simple
node T represents a chain of problems that have to be solved in order to
solve T. FD-graph algorithms may be adapted in order to be used to determine minimal representation of AND-OR graphs. In this case an interesting development may be to extend the FD-graph formalism by introducing weighted arcs in order to study heuristic strategies.

5. CONCLUSION

In this paper the problem of determining minimal coverings of hypergraphs has been studied. A graphic representation of hypergraphs has been
proposed and properties of minimal coverings have been investigated.

The computational results proved in this paper are summarized in the
following table where they are compared with related results which hold
for graphs:

TYPE OF COVERING	COMPLEXITY FOR GRAPHS	COMPLEXITY FOR HYPERGRAPHS
NONREDUNDANT SUBHYPERGRAPH	$O(m^2)$	$O(|\mathcal{H}|^2)$
MINIMUM EQUIVALENT SUBHYPERGRAPH	NP-COMPLETE (MINIMUM EQUIVALENT DIGRAPH)	NP-COMPLETE
SM-COVERING	-	$O(|\mathcal{H}|^2)$
OSM-COVERING	-	NP-COMPLETE
HM-COVERING		NP-COMPLETE
SHM-COVERING	$O(n \cdot m)$	NP-COMPLETE
O-COVERING	(TRANSITIVE REDUCTION	NP-COMPLETE

where n and m are the number of nodes and arcs in the graph and $|\mathcal{H}|$ is
the length of the representation of the hypergraph.

Besides showing how the complexity of covering problems increases when we go from graphs to hypergraphs the results provided in the paper are also devoted to determining efficient algorithms for polynomially solvable covering problems. Concerning this point by generalizing results in [2], it may be proven that in the case of SM-coverings, when hypergraphs degenerate into graphs (all source sets are singletons), the complexity of the given FD-graph algorithms coincide with the best efficiency which is known for graph algorithms.

Applications to functional dependencies in relational theory and to and-or graphs manipulation were finally sketched.

REFERENCES

[1] Aho,A.V., Garey,M.R. and Ullman,J.D., *The transitive reduction of a directed graph*. SIAM J. on Computing, 1 (1972), pp. 131-137.

[2] Ausiello,G., D'Atri,A. and Saccà,D., *Graph algorithms for functional dependency manipulation*. JACM 30, 4 (Oct. 1983), pp. 752-766.

[3] Batini,C. and D'Atri,A., *Rewriting systems as a tool for relational data base design*. LNCS 73, Springer-Verlag (1979), pp. 139-154.

[4] Berge,C., Graphs and hypergraphs. North Holland, Amsterdam (1973).

[5] Boley,H, *Directed recursive labelnode hypergraphs: a new representation language*. Artificial Intelligence 9 (1977), pp. 49-85.

[6] Fagin,R, Mendelzon,A.O. and Ullman,J.D., *A simplified universal relation assumption and its properties*. ACM TODS, 7,3 (1982), pp. 343-360.

[7] Garey,M.R. and Johnson,D.S., Computers and intractability: a guide to the theory of NP-completeness. Freeman, San Francisco (1979).

[8] Gnesi,S., Montanari,U. and Martelli,A., *Dynamic programming as graph searching: an algebraic approach.* JACM 28,4 (1981), pp. 737-751.

[9] Lipski,W., *Two NP-complete problems related to information retrieval.* Fundamentals of Computation Theory. LNCS 56, Springer-Verlag, (1977), pp. 452-458.

[10] Maier,D., *Minimum covers in the relational data base model.* JACM 27,4 (1980), pp. 664-674.

[11] Maier,D., *Descarding the universal instance assumption: preliminary results.* Proc. XP1 Conf., Stony brook, NY (1980).

[12] Maier,D. and Ullman,J.D., *Connections in acyclic hypergraphs.* 1st Symposium on Principles of Data Base Systems, Los Angeles (1982).

[13] Nilsson,N.J., Problem solving methods in artificial intelligence. McGraw Hill, New York (1971).

[14] Paz,A. and Moran,S., *NP-optimization problems and their approximation.* In Proc. 4th Int. Symp. on Automata, Languages and Programming, LNCS, Springer-Verlag, 1977.

[15] Ullman,J.D., Principles of Data Base Systems. Computer Science Press, Potomac, Md. (1980).

[16] Yannakakis,M., *A Theory of Safe Locking Policies in Database Systems.* JACM 29,3 (1982), pp. 718-740.

PART II

OPTIMAL DESIGN OF
PARALLEL COMPUTING SYSTEMS

PARALLEL COMPUTER MODELS: AN INTRODUCTION

G. Ausiello
Dipartimento di Informatica e Sistemistica
University of Roma

P. Bertolazzi
Istituto di Analisi dei Sistemi ed Informatica
CNR, Roma

SUMMARY

1. Introduction
2. Abstract models of parallel computer systems
3. Various forms of parallelism in computer systems
4. Classes of parallel machines and algorithms
5. Advantages and inherent limitations of parallel processing
6. References

ABSTRACT

Various forms of parallel processing have been realized in computer systems in the last two decades, ranging from parallelization of data processing with respect to input and output operations, to the use of higly parallel arithmetic units, to the construction of networks of tightly interconnected processors. In this introductory paper we examine various examples of abstract and real parallel machines with the aim of providing the basic concepts and discuss their fundamental characteristics. Besides we briefly discuss under what circumstances and up to what extent parallel devices may provide a more efficient solution to computational problems.

1. INTRODUCTION

The interest for parallel computation arises from various points
of view. First of all the organization of a computer system based on
more than one processor may be used to increase the throughput and the
reliability of the system; besides, in many cases, the possibility of
performing several arithmetical operations simultaneously may increase
the efficiency of a computation. These needs were present since the
beginning of the introduction of electronic computers and several multiprocessor computer systems were realized in the last two decades. The
great technological advances in microelectronics have made parallel processing much more widespread and have made it possible to build systems
with hundreds or thousands of computing units based on various organization principles. A speed up of 10 to 10^4 times in computation time
became then possible. Processing tasks whose solution would have been
extremely costly or even unfeasible on serial computers could instead
be attacked with a parallel machine. A typical example in image processing is the processing of satellite pictures: a LANDSAT picture is
made up of $8 \cdot 10^6$ pixels; when we process the image for various applications under real time constraint we may have to perform up to 10.000
operations per pixel per second which would require a machine with
computing power of 10 gips (10^3 times larger than the most efficient
serial computer would allow). Such kind of problems may hence only be
approached if we have a parallel machine (a 100 × 100 array of processors, for example).

On the other side, available multiprocessor systems would not be
useful if we would not know how to solve problems by means of parallel
algorithms. Hence the first motivation for the study of parallel algorithms is to exploit the power of parallel computation offered by the
technology.

Besides this practical reason the study of parallel algorithms is
also interesting in order to have a better understanding of the computa-

tional nature of a problem.

In fact, if for some problems we know such properties as the structure of the flow of data, the possible parallelization of computation steps, the trade-offs between different particular complexity measures or the lower bounds expressed in terms of a global complexity measure which takes into account the cost of execution, the cost of communication and the number of processors, we have a much deeper knowledge of the intrinsic computational properties of that problem.

Finally, as it has been pointed out in [Megiddo 83] the study of the parallel computer solution of one problem may turn out to be useful to provide an efficient serial solution for another related problem.

In this introductory paper we will give a presentation of various abstract parallel machine models and of the basic results that may be established for these models (§2). Then we will discuss the various forms in which parallel processing has been introduced in computer systems (§3) and provide an overview of the main classes of (real or realistic) parallel machines with a brief discussion of their fundamental characteristics (§4).

Finally (§5) we consider what are the advantages in terms of efficiency which may be obtained by using parallel processing systems and, on the other side, what are the limitations that make parallel processing convenient only for particular classes of problems.

2. ABSTRACT MODELS OF PARALLEL COMPUTER SYSTEMS

The ability of performing synchronous or asynchronous parallel steps of computation was introduced in formal systems and abstract machines since the early stages of the development of theoretical computer science, not necessarily with the aim of modeling computer systems but often in connection with the mathematical description of physical, biological, physiologycal social phenomena where concurrent evolution

of events and actions performed by several actors is not the exception
but the normal behavioural characteristic. This is the case with the
first introduction and studies on *cellular automata* which started in
the early fifties [Von Neumann 51] in connection with research on neural
networks and the brain and in connection with the mathematical theory
of self-reproduction [Moore 62]. Research on cellular automata (tes-
sellation automata, iterative arrays [Cole 64]) was continued by Myhill,
Amoroso, Yamada, Maruoka, etc. Only more recently, cellular automata
and other kinds of polyautomata were considered as models of parallel
computing machines and even as possible architectures of real parallel
computers with applications in image processing and recognition (see
[Duff, Levialdi 81]). Also in the case of *developmental systems* (or
L-systems, introduced by Lindenmayer in 1968) the original aim was to
study biological behaviour such as the growth of branches and leaves
in plants according to specific patterns or the regeneration of parts
of the body in worms etc. Then the underlying mathematical structure
was deeply studied by several authors and provided an interesting model
for parallel synchronous rewriting systems (see [Rozenberg, Salomaa 80]).

   Petri nets (introduced in Petri's thesis [Petri 62]) provide a third
example of model of parallelism which, though now widely applied in the
representation of parallel computations [Peterson 77], was not originally
intended to describe parallelism in computing, but, much more generally,
was motivated by the ambition of modeling the flow of information in
systems (e.g. physical and human organizations) in which events may
occur concurrently and asynchronously, with limitations due to some constraints (such as precedence, mutual exclusion etc.).

   More closely related to the modelisation of parallel computers, has
been the introduction of parallelism in classical abstract machine models
such as *Turing machines* (see for example [Kozen 76; Chandra, Stockmeyer
76] or *random access machines* (see for example [Pratt, Stockmeyer 76;
Savitch, Stimson 76]). In these cases the main objective was to study

the computational power of such systems by comparing their resource
bounded complexity classes. An example of a result that can be proved
for some parallel abstract machines and that in general remains a conjecture, though supported by strong evidence, is that if we allow an
unbounded amount of parallelism the class of functions computable in
polynomial time by a parallel machine corresponds to the class of functions computable by a serial machine which makes use of a polynomial
amount of storage. In particular this result holds for the model of
SIMD machine (see next paragraph) with a shared global memory, called
SIMDAG in [Goldschlager 78], where all parallel processing units (PPU)
are RAMs with the usual RAM instruction set augmented by parallel instructions which are broadcast by the CPU and executed simultaneously
by the PPUs. The quest for a general universal model of parallel abstract machine, capable of simulating all known abstract and real parallel computers, thereby providing a sort of parallel version of
Church's Thesis[*], and the need of establishing meaningful comparisons
of computational power among them, has brought to the characterization
of various general classes of parallel machines.

   The more general class, called *paracomputers* in [Schwartz 80] or
idealistic parallel machines correspond to the following model: N
identical processors share a common memory which they can read and write
simultaneously in a single cycle (Fig. 1). This model is clearly unreali-
stic due to physical fan in limitations. It can only provide a first
approximate idea of the possible parallel solution of a problem and it

---

(*) Church's Thesis may not be trivially extended to parallel machine
    models because in presence of nondeterminacy parallel computations
    may allow more than one output. In fact the sets of relations com-
    puted by parallel programs with non determinacy may not be even
    semidecidable [Chandra 79].

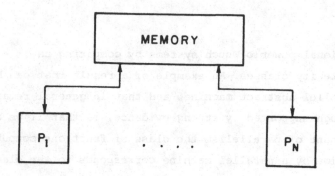

Fig. 1. The paracomputer

can be used to derive trivial upper and lower bounds for parallel time
complexity: in fact we know that if an algorithm requires time $O(t(n))$
on a serial computer in the best case it can be executed in time
$O(t(n)/N)$ on a paracomputer and, on the other side, if a problem requires time $\Omega(t(n))$ on a serial computer its lower bound on a parallel
machine will be at least $\Omega(t(n)/N)$.

The second class is the class of *realistic* parallel computers,
[Galil, Paul 81; Valiant, Brebner 81] called *ultracomputers* in [Schwartz
80], which are based on the following model: N identical processors are
located in the nodes of a potentially infinite recursive graph structure;
all processors are connected, along the edges of the graph, to a small
number of neighbours (for example a constant number d or a slowly growing number $\log_2 N$). This limitation of fan -in makes the model more

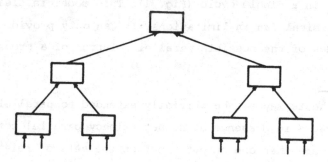

Fig. 2. The tree machine

realistic but clearly, at least in general, less efficient than a pa-

racomputer. In the case that the processors are restricted to be finite
state automata we obtain a smaller class of machines (which is still
fairly general and includes for example iterative arrays and tree structures (Fig. 2), called *conglomerates* and claimed in [Goldschlager 78] to
include all synchronous parallel machines which could be feasibly built.

In more general cases the processors may be assumed to be RAMs with
a constant number of registers (possibly of bounded capacity). The main
results which have been proved both for conglomerates and for the more
general machine models was the existence of universal interconnection
patterns. For example, in the case when the processors are RAMs it can
be shown that there exists a universal parallel machine U such that given
a parallel machine C with p processors which operates on input x of
length n in time t, U simulates C on input x in time $0(t \log^2 p)$ by using
not more than $0(p)$ processors ·

Further research in this direction has been developped with the
aim of finding efficient simulations of abstract models on more realistic parallel architectures. (See [Vishkin 83] for a survey). The simulation is realized by implementing parallel algorithms designed for the
abstract model on the realistic machine. It is shown that in general
small increase of the parallel complexity is obtained [Vishkin 82],
[Eckstein 79],[Schwartz 80].

A central role in abstract parallel machine models such as in real
parallel computers is played by interconnection schemes and communication problems. In order to perform an efficient computation it may be
required that the largest distance between two processors in a network
be limited to a slowly growing function of the total number of processors,
say $\log_2 N$.

While simple planar structures of N processors such as rectangular
and hexagonal arrays determine a $\sqrt{N}$ growth of interprocessor communication time, the logarithmic distance is realized in structures such as
the k dimensional cube, the shuffle - exchange network (Fig. 3), the mesh
of trees [Leighton 81].

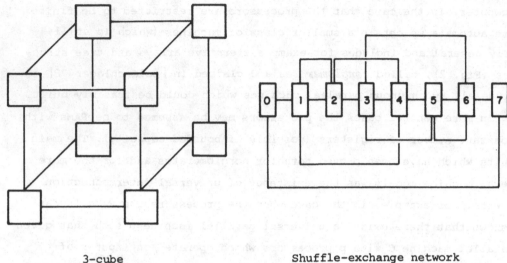

3-cube   Shuffle-exchange network

Fig. 3.

Unfortunately the practical application of these interconnection schemes
is limited by three factors:

- the difficulty of performing interprocessor communication in logarithmic
  time due to constraints on the capacity of communication lines;
- the fan-in physical limitations which require that the number of
  neighbour processors be either constant or at most logarithmic;
- the wiring constraints, which do not allow, especially for VLSI implementations, more than two or three levels of wiring and which pose
  restrictions on the density and length of connection wires in a layout.

Concerning the first problem, a remarkable result in [Valiant,
Brebner 81] shows that a randomized routing algorithm may guarantee a
logarithmic time communication among processors in various structures
such as the k-cube, the shuffle exchange network etc.

The second problem may be overcome by adopting the cube connected
cycles (CCC) interconnection scheme [Preparata, Vuillemin 81] where
every processor is constantly connected only to three neighbours, still
preserving the general properties of the k-cube architecture (Fig. 4).

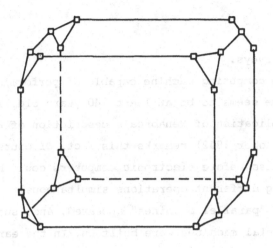

3-CCC

Fig. 4.

The CCC has been shown to be optimal for several problems with respect to the area x time^2 complexity measure for VLSI implementations. Nevertheless also in this case such as in most non-planar networks a severe limitation to the physical realization comes from the technological problems connected with the layout.

3. VARIOUS FORMS OF PARALLELISM IN COMPUTER SYSTEMS

Let us now consider how parallelism has been introduced in real computer systems and how it gave rise to various kinds of parallel computers and multiprocessors. The history of the evolution of parallel computer systems has been extensively discussed in several survey papers, together with various approaches to the characterization of such systems and to the classification of the corresponding algorithms [Flynn 66; Baer 73; Kuck 77; Ramamoorthy, Li 77; Enslow 77; Reddy, Hon 79; Kung 80]. Here and in the following paragraph we limit ourselves to providing some examples of various kinds of parallel machines and a discussion of the characteristics of the most important classes of parallel systems which have been introduced in the literature. References concerning the particular machines which we are taking into consideration may be found in the above cited surveys.

The idea of a computing machine capable of performing more than one operation at a time seems to be at least 140 years old. In [Kuck 77] a reference to a publication of Menabrea's description of Babbage's lectures in Turin (october 1842) remarks this fact. Of course more then 100 years had to pass before electronic computers could be built capable of performing different operations simultaneously. From this point on many designs of "parallel machines" appeared, and, successively, prototypes and commercial machines were built on. In the early 50s the first prototypes appeared, such as the multiprocessor Model V of the Bell Telephone Laboratories, with two processors, and the multioperation processor of Leondes and Rubinoff oriented toward a drum memory. Successively, in the 60s, many multioperation machines appeared. First of all, in that period, most of the computers had undergone a transformation, oriented toward increasing the throughput, and more than one processor with different functional utilizations (I/O and processing) had been introduced in a computer, connected via multiple bus systems. Gradually parallelism between decoding and execution of operations and between execution of different arithmetic operations was introduced (IBM 360/91, CDC 6600, CDC 7600) by allowing several functional units to perform specific arithmetic functions in parallel. In some cases even several general purpose processing units were coupled in order to provide a higher efficiency (IBM 360/67, UNIVAC 1110 etc.).

More massive use of several processing units was introduced in the late 60's and in the 70's in the realization of the first, so called, array processors (such as ILLIAC III and ILLIAC IV, the last one with 64 processing elements each with a small private memory) or various other kinds of systems oriented toward the fast parallel processing of vectors (such as the SDC PEPE, 1971, the TEXAS ASC, 1972, the CDC STAR-100, 1973, the CRAY-1, 1976, down to the more recent vector processors IBM 3838 and Hitachi IAP). Other examples of parallel computers appeared in connection with the fast processing of picture cells in images (such as CLIP, a

96 × 96 array of processors, or MPP, the Massively Parallel Processor).

Beside these examples of systems with a tight coupling among processors several other kinds of multiprocessor architectures have been
developped during the 70 such as multimini/multimicroprocessors (closely
coupled systems, such as DAP,C.mmp and Cm* with a number of processor
ranging from 16 to 256 in various stages of development) or local networks (losely coupled systems).

If one goes deeply in the hardware and software organization of
these machines, one can easily see that the term "parallel processing"
has been used in many different ways. In fact twenty five years ago it
referred to arithmetic operations on whole words rather than on one bit
at a time.

Also the parallel execution of instructions of programs and I/O
operations in multiprogramming has been seen as parallel processing.

The first true parallel machines can be thought of to be the
machines of the 60's. They are often called array processors; the name
is due to the fact that these machines can operate with high performance
on arrays of data. These machines have a new form of parallelism: they
operate simultaneously on different elements of the same vector. Two
basic principles of organization of parallel operation are present in
the computer systems described until now: the pipelined organization and
the single instruction multiple data (SIMD [Flynn 66]) organization. In
the pipelined machines such as CDC STAR-100 mentioned above a computational process is segmented into several different subprocesses, which
are executed by dedicated autonomous units. Successive processes can be
overlapped, analogously to an industrial assembly line [Ramamoorthy 77].
In the SIMD machines, such as ILLIAC, all the units execute the same
computational process on different data under the control of a central
processing unit.

Successively, other kinds of parallelism were developed: on one
hand the parallel asynchronous execution of different tasks of the same

job on the processors of a multiprocessor system; each task requires a
great amount of computation and the communication can be performed via
global variables in a shared memory or via messages sent through high
speed lines.

These multiprocessor systems are often called MIMD (Multiple Instruction Multiple Data stream): Cm* and DAP are examples of this kind
of machines. On the other hand the distribution of small amount of
computation among simple processing units, connected together with
simple geometrical architectures has been made possible by the evolution
of VLSI. Many processing units emebedded in a chip alternatively perform simple computations and send data, synchronized by a clock; the
input data are entered by a driver and "pipelined" in the circuit, while
the "instructions" are realized by the components of the circuit itself.
The name "systolic" is used to refer to this way of parallel processing.

4. CLASSES OF PARALLEL MACHINES AND ALGORITHMS

Among all possible parameters which may be used to characterize
parallel computer systems and which give rise to such a wide .variety
of architectures, as we have seen in the preceding paragraphs, the following appear to be the most relevant:

a) *Quality of processors*: as we have seen, the processors which operate
   in parallel may be
   - homogeneous
   - non homogeneous
   and, in the first case, they may be
   - functionally specialized units (e.g. floating point adders and
     multipliers)
   - general purpose processors.

b) *Control of concurrent operations*: three main kinds of concurrency

control schemes appear in the systems which are under discussion
[Kung 77]

- centralized control: all processing units are synchronized under
  the supervision of a central unit,
- distributed control: in this case all units may operate either
  synchronously (via a clock) or asynchronously (via messages),
- control via shared data: under this kind of asynchronous control,
  processes activate each other by means of global variables.

c) *Geometry*[*] *of the interconnection scheme*: whatever is the characterization of a parallel machine according to the first parameters, a
large choice of regular communication geometries may be used: one
dimensional array, binary tree, planar grid, cube, shuffle etc. In
some cases, mainly in connection with asynchronous control schemes,
irregular geometries may also be adopted.

Actually it has been observed by several authors that the basic
operation principles of most real parallel computer systems fall into
just a few classes with respect to those which may arise by combining
the said parameters in all possible ways.

On one side we have machines with a synchronous mode of operation;
among them we may distinguish the following classes:

i) *SIMD* (Single Instruction Multiple Data stream). In this case
homogeneous processors, organized in a regular network, all perform the same operation, broadcast by the central processing unit,
at the same time (see for example ILLIAC IV).

ii) *Pipeline*. As we have already observed, in this mode of operation
data flow in the network of processors (which perform on them
specialized functions) such as products in an assembly line. In

---

(*) Since the interconnection scheme has to be embedded in a "metric"
space the mere topology does not provide a sufficient characterization.

some examples of pipeline machines such as CRAY-1 or STAR-100 we
have multiple stages floating point adders and multipliers. In
other cases (vector processors) we may have elements of two
vectors flowing through multipliers and adders in order to perform
sequences of scalar products.

iii) *Systolic*. This type of synchronous, distributed control organization derives its name from the "systoles", the rythmic contrac-
tions of the heart which make the blood flow in the arteries. In
this case each processor organized in a multidimensional network
takes data from nearby processors, performs a short computation
and sends data again to nearby processors. Typically systolic
systems may be realized using VLSI technology. In some applications
systolic and pipeline modes of operation may be combined in order
to increase the efficiency over sequences of computations.

On the other side we have asynchronous multiprocessors; in this
case there is mainly one mode of operation:

- MIMD (Multiple Instruction Multiple Data stream): various processors (usually general purpose processors connected by crossbar
switches or high speed buses) with independent instruction counters
perform different operations on different data. Communication and
cooperation between processors is realized via shared variables
or via messages (see for example Cm* cr Pluribus).

From the point of view of the algorithms which are more suitable
to be executed by parallel machines the given classification allows us
to determine a first rough distinction: on one side we have algorithms
in which the amount of computation which may be performed by every processor autonomously, without the need for an interprocessor communication step, is large (in this case we speak of *large module granularity*
[Kung 80] while on the other side we have frequent communication steps
and very short processing steps (*small module granularity*). The first

kind of algorithms will be better suited for an MIMD machine with asynchronous concurrency control. In this case the user language needed for programming the algorithm will have to be rich enough to provide visible interprocesses communication constructs for the specification of a logical level and the system language will have to support high level communication and synchronization primitives.

Typical examples of this kind of algorithms are concurrent data base management and relaxed global and local optimization.

Algorithms with small granularity are instead suitable for synchronous machines. The overhead due to synchronization and frequent communication would be unbearable on an MIMD machine. In this case a hardware direct data communication path has to be provided and the programming language constructs, which are needed, may be much more simple at user level and much more related to the physical architecture of the processors than to the logical organization of processes. Examples of this kind of algorithms will be referred in the next paragraph.

5. ADVANTAGES AND INHERENT LIMITATIONS OF PARALLEL PROCESSING

Parallel algorithms were studied since 1960 (see a survey in [Miranker 1971]), although the first parallel machines were built only some years later.

The advent of multiprocessors before and the recent advances in VLSI technology provided impetus to the investigation of parallel algorithms for different kinds of problems. In the field of numerical linear algebra (see [Heller 79] for a survey), parallel algorithms were studied for the solution of general and special linear systems of equations, computation of eigenvalues, evaluation of arithmetic expressions, operations on matrices (product, inversion), FFT etc. In nonnumerical calculus, parallel sort, merge and search have received great attention. Parallel algorithms for operations on particular data

structures (priority queues, graphs) and on data bases, have been suggested [Kung , Lehman 80, Munro 79]; finally parallel algorithms for
combinatorial optimization problems have been proposed (graph problems
as max flow and connected components,   , matching, scheduling,etc.).

In [Heller 78],[Kung 80],[Schwartz 80] [Vishkin 83],[Kindervater
83] very wide bibliographies on parallel algorithms are contained, in
[Kook 83] a classification of problems in terms of parallel complexity
is given.

The advantages of parallel processing are evident in many cases.
Serial algorithms, for which linear time is required, can be performed
in logarithmic time on parallel machines with a linear number of processors: examples are the evaluation of general expressions, the inner
product, the addition of N values [Heller 78],[Schwartz 80].

The sort of N elements can be performed in $O(\log^2 N)$ on N processors
connected with the shuffle network [Schwartz 80] using bitonic sort
(which would require $O(N \log^2 N)$ on a serial machine) while the same
algorithm requires $O(\sqrt{N})$ on a two dimensional $\sqrt{N} \times \sqrt{N}$ array [Thompson
77]. The classical matrix product algorithm requires $O(\log N)$ on the
cube connected computer with N^3 processors, and $O(N)$ on an array of
$O(N^2)$ hex-connected processors [Dekel 80],[Kung 80].

Another field of application is global optimization: both probabilistic and deterministic (gradient technique, search technique) serial
methods require very high computational efforts in evaluating the function; this characteristic makes the use of parallel processing methods
attractive, as much of the computation may be carried out as a group of
parallel tasks [Mc Keown 80],[Dixon 81]. Looking at the examples, however, one can observe that the bounds defined in §2 in the case of the
ideal model of paracomputer are very rarely achieved; moreover, in
general, the performance depends on the computational model chosen. This
is due firstly to the intrinsic characteristic of the problem:

- problems, which require exponential time on a serial machine, can not be solved in polynomial time unless using an exponential number of processors (see [Chi-Chin Yao 81] for the knapsack problem);

- problems with N inputs and one output can not be solved, in a parallel system with N processors, in less than log N steps, if only unary and binary operations are admitted [Heller 78] (see the addition of N numbers), even if the paracomputer model is adopted.

The other fundamental reason of this limit is the fact that, while in the serial algorithm only computation steps must be evaluated, in parallel algorithms one must keep into account the overhead due to communication steps [Lint 81]. A quantitative limit is given in [Gentleman 78] on the performance of the matrix product in a N × N array processor. This limit can be generalized to the ultracomputer model: it is impossible to solve a problem on an ultracomputer in less than $O(D)$ steps (D is the maximum distance between two processors) if the input data are required to be moved to any processor of the system. The odd even transposition sort of N elements requires $O(N)$ in a linear array with N processors [Kung 80]; the bitonic sort of N^2 elements requires $O(N)$ in a N × N array processor, as $O(N)$ communication steps are required to move data to the farthest position [Thompson 77]; the same is for the product of N × N matrices [Kung 80].

In systems communicating via a shared memory, memory contention causes the reduction of the performance: in this case more advantages can be obtained for problems in which computation time is much greater than communication time (see dynamic programming [Al Dabass 80] and numerical optimization [Dixon 81; Mc Keown 80]).

In VLSI fan in and lay out of wires limit the connection geometries, and the communication between distant processors strongly influence the performance. Of course there are problems for which such limit can be overcome: for some problems in image processing, and for particular classes of dynamic programming problems it is possible to detect "locality" in the operations, which make the number of communication steps independent from the architecture and the input size.

Finally one can see that the efficiency achievable with parallel processing for a particular problem is strongly influenced by the parallel machine chosen (MIMD, SIMD, systolic) and in the VLSI technology, by the communication geometry: in fact the data of a problem must be moved according to a specific communication pattern (this is why, for example, sort on a bidimensional array is less efficient than in a shuffle exchange network). For this reason it would be greatly helpful to have computational models capable of evidentiating the intrinsic parallelism of a problem and its data communication pattern. In this way it would be possible to obtain indications on the convenience of studying a parallel algorithm for the given problem and help in the design of the best architecture for its execution. This research aim has been partially tackled in VLSI studies where the area $\times$ time^2 bound may be considered a way of measuring how good is the matching between algorithm and circuit [Thompson 79]. The computational model proposed by Thompson is discussed in [Chazelle 81] and [Bilardi 81] who propose to restrict the hypothesis on independence of communication time on the wire length obtaining new values for the lower bounds AT^2. Most of the theoretical research on parallel computation is likely to be directed toward this goal in the future years.

6. REFERENCES

[Baer 73]   J.L. Baer: A survey od some theoretical aspects of multiprocessing. ACM Computing surveys, vol. 5, n.4, March 1973.

[Bilardi 81] G. Bilardi, M. Pracchi, F.P. Preparata: A critique and appraisal of VLSI models of computation. Manuscript, 1981.

[Chandra 79] A.K. Chandra: Computable nondeterministic functions, Proc. of 4th IBM Symposium on Math. Found. of Comp. Sc., Tokyo, 1979.

[Chandra, Stockmeyer 76]   A.K. Chandra, L.J. Stockmeyer: Alternation,
             Proc. 17th FOCS, 1976.

[Chazelle 81]   B. Chazelle, L. Monier: A model of computation for VLSI
             with related complexity results. Carnegie Mell
             University, Tech. Rep. n. CMU-CS-81-107, Feb. 1981.

[Chi-Chin Yao 81]   A. Chi-Chin Yao: On the parallel computation for the
             knapsack problem, 13th Symposium of Theory of Comp.,
             Milwaukee 1981.

[Cole 64]   S.N. Cole: Real time computation by iterative arrays of
             finite state machines, Doctoral Thesis Harvard University,
             Cambridge, Mass., 1964.

[Cook 83]   S.A. Cook: The classification of problems which have fast
             parallel algorithms. Found. of Comp. Theory Borgholm,
             Sweden Aug. 1983, Springer Verlag Ed.

[Dekel 80]   E. Dekel, D. Nassimi, S. Sahni: Parallel matrix and graph
             algorithms. From 18th Allerton Conference on Communica-
             tion Control and Comp. Oct. 79.

[Dixon 81]   L.C.W. Dixon: The place of parallel computation in numerical
             optimization, CREST-CNR Summer School on design of
             numerical algorithms for parallel processing, Bergamo
             June 1981.

[Duff, Levialdi 81]   M.J.B. Duff, S. Levialdi edits: Languages and
             architectures for image processing, Academic Press, 1981.

[Eckstein 79]   D.M. Eckstein: Simultaneous memory access, TR-79-6,
             Computer Sc. Dep., Iowa State University, Arnes, Iowa
             1979.

[Enslow 77]   P.H. Enslow: Multiprocessor organization: a survey, ACM
             Comp. Surveys, vol. 9, n. 1, March 1977.

[Flynn 66]   M.J. Flynn: Very high speed computing systems, Proc. IEEE, 54, Dec. 1966.

[Galil, Paul 81]   Z. Galil, W.J. Paul: An efficient general purpose parallel computer, 13th STOC, 1981.

[Gentleman 76]   W.M. Gentleman: Some complexity results for matrix computations on parallel processors, Journal of ACM, vol. 23, n. 1, Jan 1976.

[Goldschlager 78]   L.M. Goldschlager: A unified approach to models of synchronous parallel machines, Proc. 10th STOC, 1978.

[Goldschlager 82]   L.M. Goldschlager: A universal interconnection pattern for parallel computers. J. of ACM, vol. 29, n. 4, Oct. 1982.

[Heller 78]   D. Heller: A survey of parallel algorithms in numerical linear algebra, Siam Review, vol. 20, n. 4, Oct. 1978.

[Kindervater 83]   G.A.P. Kindervater, J.K. Lenstra: Parallel algorithms in combinatorial optimization: an annotated bibliography. Mathematisch Centrum Techn. Rep. n. BW 189/83, Aug. 1983.

[Kozen 76]   D. Kozen: On parallelism in Turing machines, Proc. 17th FOCS, 1976.

[Kuck 77]   D.J. Kuck: A survey of parallel machine organization and programming, ACM Comp. Surveys, vol. 9, n. 1, March 1977.

[Kung 80]   H.T. Kung: The structure of parallel algorithms in Advances in Computers, vol. 19, ed. by Marshall C. Yovits, Academic Press, 1980.

[Kung, Lehman 80]   H.T. Kung, P. Lehman: Systolic (VLSI) array for relational data base operations, 1980 ACM SIGMOD International Conference on Management of Data, Los Angeles, May 1980.

[Lint 81]   B. Lint, T. Agerwala: Communication issues in the design and
            analysis of parallel algorithms, IEEE Trans. on Softw.
            Eng., vol. 7, n. 2, March 1981.

[Mc Keown 80]   J.J. Mc Keown: Aspects of parallel computation in numeric-
            al optimization, on Numerical Techniques for Stochastic
            systems. F. Archetti, M. Cugiani eds., North Holland
            Pub. 1980.

[Megiddo 83]   N. Megiddo: Applying parallel computation algorithms in
            the design of serial algorithms, Journal of ACM, 30,4,
            1983.

[Miranker 71]   W.L. Miranker: A survey of parallelism in numerical ana-
            lysis, SIAM Review, 13, 1971.

[Moore 62]   E.F. Moore: Machine models of self-reproduction, Proc. Symp.
            Appl. Math., 14, 1962.

[Munro 79]   J.I. Munro, E.L. Robertson: Parallel algorithms and serial
            data structures, 17th Annual Allerton Conference on
            Communication, Control and Computing, Oct. 1979.

[Peterson 77]   J.L. Peterson: Petri nets, ACM Comp. Surv. 9,3, 1977.

[Petri 62]   C.A. Petri: Kommunication mit Automaten, Schrift des RW
            Inst. f. Instr. Math. an der U. Bonn Heft 2, Bonn, 1962.

[Pratt, Stockmeyer 76]   V.R. Pratt, L.J. Stockmeyer: A characterization
            of the power of vector machines, JCSS, 1976.

[Preparata, Vuillemin 81]   F.P. Preparata, J. Vuillemin: The cube-con-
            nected cycles: A versatile network for parallel computa-
            tion, Communications of ACM, 24, 5, 1981.

[Ramamoorthy, Li 77]   C.V. Ramamoorthy, H.F. Li: Pipeline architecture,
            ACM Comp. Surveys, vol. 9, n. 1, March 1977.

[Reddy, Hon 79]   D.R. Reddy, R.W. Hon: Computer architecture for vision,
                  Computer vision and sensor based robots, ed. G.G. Dodd,
                  L. Rossol, Plenum Press, New York, 1979.

[Rozenberg, Salomaa 80]   G. Rozenberg, A. Salomaa: The mathematical
                  theory of L-systems, Academic Press, 1980.

[Savitch, Stimson 76]   W.J. Savitch, M.J. Stimson: Time bounded random
                  access machines with parallel processing, TR IW 67/76,
                  Math. Centrum, Amsterdam, 1976.

[Schwartz 80]   J.T. Schwartz: Ultracomputers, ACM TOPLAS, 2,4, 1980.

[Thompson 79]   C.D. Thompson: Area time complexity for VLSI, Proc. of
                  the 11th Annual ACM Symp. on the Theory of Comp. May
                  1979.

[Thompson 77]   C.D. Thompson, H.T. Kung: Sorting on a mesh connected
                  parallel computer, Comm. of ACM, vol. 20, n. 4, Apr.
                  1977.

[Valiant, Brebner 81]   L.G. Valiant, G.J. Brebner: Universal schemes
                  for parallel communication, 13th STOC, 1981.

[Vishkin 82]   U. Vishkin: Parallel-Design space, Distributed-Implementa-
                  tion space (PDDI) general purpose computer. RC 9541,
                  IBM T.J. Watson Research Center, Yorktown Heights, 1982.

[Vishkin 83]   U. Vishkin: Synchronous Parallel Computation. A survey,
                  1982 Manuscript.

[Von Neumann 51]   J. Von Neumann: The general and logical theory of
                  automata, in Cerebral Mechanisms in Behaviour, Hixon
                  Symposium, 1948 (Wiley, N.Y. 1951).

FUNCTIONAL ORGANIZATION OF MIMD MACHINES

G. Cioffi
Dipartimento di Informatica e Sistemistica
Università di Roma, La Sapienza

1. INTRODUCTION

In the computer science community there is not a complete agreement on what a multiprocessor system is.

There is a restricted definition that sounds:

a multiprocessor architecture is one that consists of at least two processors satisfying the following conditions:

i)   the processors share a global memory
ii) they are not highly specialised
iii) each processor is capable of doing significant computation.

Another definition, which enlarges the class of multiprocessor architectures, is based on the concept of <u>instruction stream</u> and <u>data stream</u>. A computer executes a sequence of instructions on a sequence of data: multiplicities in these streams lead to four classes of computer architectures:

SISD   Single Instruction Single Data
SIMD   Single Instruction Multiple Data
MISD   Multiple Instruction Single Data
MIMD   Multiple Instruction Multiple Data

A multiprocessor architecture falls in the class of MIMD machines and can be defined as follows:

a multiprocessor architecture is one that consists of at least two computers/processors which cooperate to execute multiprocessing.

In the sequel we will adopt this second definition which is more general and comprehend the class of multiprocessors with shared global memory and the class of multiple processors/computers which cooperate exchanging messages each other via parallel or serial communication links.

At this point a question naturally arises: why multiprocessing? There are many motivations that can be used for answering to this question:

i)   the revolution of microelectronics and VLSI offers more and more powerfull microcomputers/microprocessors on chip at extremely low price which can be used as building blocks of multiprocessors with better performance/cost index
ii) multiprocessing reduces the computation time exploiting the intrinsic parallelism of the application
iii) modular architectures, tipical of multiprocessor systems, lead to graceful growth and degradation
iv) VLSI modules offer greater reliability, omogeneous modular architectures, greater availability.

2. TIGHTLY COUPLED SYSTEMS

A tightly coupled MIMD machine corresponds to the first given definition of multiprocessor system and its HW/SW organization can varies according to the following charecteristics:

- symmetric structure
- semisymmetric structure

A symmetric structure in its most general form is depicted in fig.1.

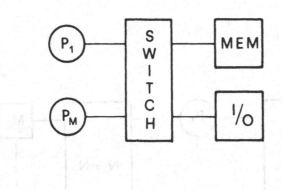

Fig.1.   Symmetric multiprocessor.

The structure in fig.1 is HW/SW symmetric if the following conditions are satisfied:

1)   All processor modules are identical
2)   Each processor has access to the whole memory
3)   Each processor has access to all peripherals
4)   Each processor is anonymous and is considered as a resource by a
     single central operating system.

The advantages of this architecture relies in the fact that the application processes ignore the architecture and the number of the processors. The operating system provides on a dynamic basis to assign a
free processor (resource) to a ready to run process. The system can
grow and degrade with complete transparency for the application software.

The concurrent processes can communicate each other on the basis of
global variables located in the common memory, or by means of message
passing technique via logical channels created by the kernel in the common memory.

Although this model of architecture seems a valid approach to achieving almost unlimited improvements in performance adding more processors,
the reality is quite different.

If the block SWITCH of fig.1 is a system bus which all units (Processor, memory, I/O modules) are attached to, it becomes a tremendous bottleneck with very rapid saturation. Suppose, for example, that τ is the

average time to execute an instruction (in μsec), and $x\tau$ (x<1) the fraction of time that a processor uses the system bus to access the memory: the number of instruction per second (MIPS) is $1/\tau$ for one processor and $1/x \cdot \tau$[1] for the system if $N \geq 1/x$. Considering that x lies in the range 0.3÷0.5, no more than two or three processors can work simultaneously. To overcome this saturation effect it is necessary to design the switch, the memory and the peripherals in a more complex way as reported in fig.2[2].

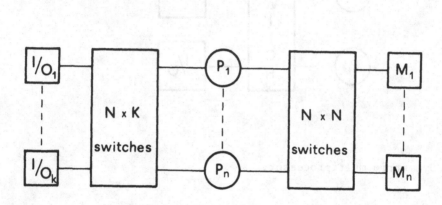

Fig.2.   Cross bar-interconnection for multiprocessors.

In this architecture the conflict in accessing memory and I/O peripherals is drastically reduced but the interconnection structure is very expensive and unreliable.

Moreover the partitioning of data structures and op code in the memory modules is a complex task and can influence greatly the performance

---

(1) This value supposes that the arbitration time to access the bus is negligible or is incorporated in $x\tau$ and moreover that the processors will synchronize themselves in utilizing the $1/x$ time slots offered on the system bus.

(2) The architecture of fig.2 has been adopted, with some modifications, in the multiprocessor system C.mmp, developed at Carnegie Mellon University during the '70's.

of the system.

Another complex problem to solve is the routing among I/O devices and processors: the communication processor-memory is always performed via a master-slave procedure, with the processor as master, whereas the processor-device communication can be activated by the devices too by means of interrupts. Considering that the processors are anonymous the routing of the interrupts to the processors is a complex task.

A semisymmetric structure for memory-coupled multiprocessor systems is a compromise to reduce the saturation effects on the system bus. The system memory is partitioned into private blocks and a common block as outlined in fig.3.

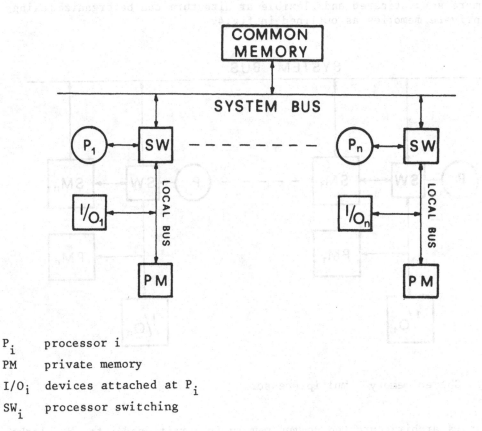

P$_i$ processor i

PM   private memory

I/O$_i$ devices attached at P$_i$

SW$_i$ processor switching

Fig.3. Global memory multiprocessor.

To avoid saturation effects on the system bus, most code must be located in private memories. The common memory is dedicated to store kernel code and communication channels or global variables.

In this way the bottleneck of the system bus is reduced but the processes must be allocated statically, at compilation time, into the processors.

The structure of fig.3 has the advantage of hardware simplicity but the rigid allocation of processes and I/O devices to the processors leads to a loss of flexibility and fault tolerance. The last point is particularly important considering that one of the reasons claimed out for introducing multiprocessor architectures, is their aptitude to tolerate processor faults.

A more sophisticated and flexible architecture can be organized using semi-private memories as outlined in fig.4.

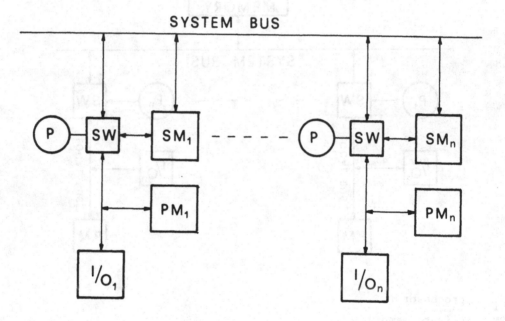

Fig.4. Shared memory   multiprocessor.

In this architecture the common memory is partitioned into N blocks, one per processor, and each block, called SM, is a dual port memory.

This architecture supports a large variety of system configurations, including memory common to all processors, memory common to some processors and private memories for each processor. The system bus bottleneck is further reduced considering that every processor accesses directly the common memory block located into its node.

3. LOOSELY COUPLED SYSTEMS

Although the concept of tightly coupled systems seems a valid approach
to achieving improvements in performance adding more processors, there
are several limitations on the diffusion of such architectures. The main
limitations are:

1) The direct sharing of memory and I/O results in access conflicts
   and saturation effects
2) Any inefficiency in the operating system is greatly amplified in
   a tightly coupled system
3) Concurrent programming languages that support effectively memory
   coupled architectures have not been adequately developed
4) A local fault can influence the entire system due to error propagation
5) The hardware complexity of a tightly-coupled system is high and
   its modularity (growth aptitude) is limited
6) Due to the hardware complexity and the operating system cruciality,
   the reliability of these systems is not very high.

For all these motivations the attention of many researches has been
focused on loosely coupled systems as an alternative more effective approach to multiprocessing.

Loosely coupled systems are multiple computer systems in which the
individual processors communicate one-another at the input-output level.
There is no direct sharing of primary memory, therefore the operating
system must be distributed with decentralized control, and the cooperation among processes must be performed via explicit messages from the
source process to the destination one. This fact implies, of course,
that both sender and receiving processors cooperate in the message exchange,whereas in memory coupled systems the receiving processor does
not partecipate in that activity.

Loosely coupled systems are often divided in two categories according
to the kind of node interconnection.

1) Multiple-computer systems if the interconnection network has high
   bandwidth and the nodes of the system are physically close each
   other, may be in the same cabinet
2) Local area network systems if the interconnection network is a
   data link with moderate bandwidth and the nodes of the system are
   physically remote each ofter.

This difference in the interconnection network determines a substantial difference in the operational of the two classes of loosely systems.
Local networks generally have the main function of sharing expensive
resources among nodes (mass storage, line printers, etc.) but do not
cooperate extensively each other; therefore local networks are not considered actually MIMD machines and will not be considered in the sequel.
Multiple computer systems have been investigate extensively from many
points of view including languages, operating systems and fault toler-

ance. In this paper we will focuse the attention on the network interconnection structures and the type of message passing techniques employ-
able in loosely coupled multiprocessor systems.

As concluding remarks we can state:

i)   memory coupled systems are particularly effective in process co-
     operation when a centralized operating system is adopted and shared
     data structures are used. The interconnection network must be ex-
     pensive otherwise becomes a bottleneck for the performance

ii) I/O coupled systems offer a greater degree of modularity and fault
     tolerance; the need of a decentralized or distributed operating sys-
     tem leeds to simpler and more clear process cooperation. The in-
     terconnection network is used only for message passing.

4. LOOSELY COUPLED SYSTEMS INTERCONNECTION STRUCTURES

There are many connection structures for linking the nodes (processors) of a multicomputer network and more will be proposed in the future.
As was stated in the previous chapter, one of the advantages of loosely
coupled systems is their aptitude to grow, that is to increase the number of nodes of the system. Therefore a useful index for comparing the
interconnection structures proposed is one that incorporates several key
factors:

i)   total interconnection cost C_T

ii) message traffic density T , supported by links or nodes

iii) message routing delay D

We will discuss some of the most interesting interconnection structures and introduce a global index as an attempt of evaluating them.

4.1. Simple interconnection structures

When the number of nodes of the architecture is not high simple topologies can be adopted to build a loosely-coupled MIMD machine. The
most popular ones are:

i)   Global bus system

ii) Star

iii) Ring

iv) Fully connected multicomputer

v)   Tree network

Global bus system

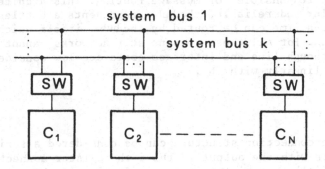

Fig.5.   Global bus system.

The general architecture is shown in fig.5. There are N computers
connected one another via K shared busses (K =1.2...N).
The number of busses can be increased for reducing traffic density
on each bus and for increasing availability as well. The number of nodes
cannot increase more than some tens because electrical reasons limit the
number of interfaces on the same bus.

Star

Star network is quite common for the simplicity of control (fig.6)
and many realizations of this architecture have been appeared.

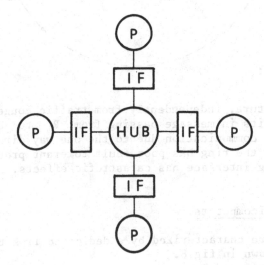

Fig.6.   Star network.

The central hub is a complex switcher supported by a dedicated computer which is responsible of message routing. This architecture is
the less modular and reliable, the hub represents a bottleneck since
one message at a time can be routed by the hub itself which is not a
crossbar switch. For all these reasons star networks cannot grow to high
N values, although line and interface costs are low considered that
they increase linearly with N .

Ring

A ring interconnection structure can be considered a series of N
shift registers with the output of the i-th register connected to the
input of the (i+1)-th register in a loop fashion as shown in fig.7.

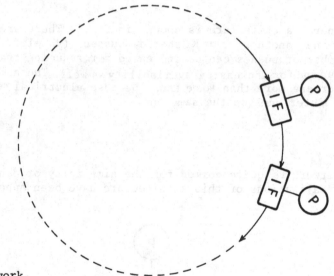

Fig.7. Ring network.

In the ring structure, independently from traffic congestion, there
is a delay of h units in message passing from P_i to P_{i+h} and N-h
units in the inverse communication due to the one way direction of message flow. Note that the ring has poor fault tolerant properties since
a failure in one ring interface has catastrofic effects.

Fully connected multicomputers

These networks are characterized by a dedicated link between each
pair of nodes, as shown in fig.8.

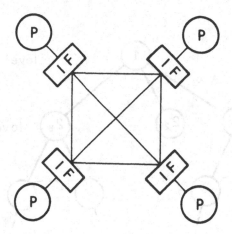

Fig.8. Fully connected net.

The traffic density on a dedicated link decreases linearly with N ,
whereas the interconnection cost is very high since the links grow with
1/2 N(N-1) and the interface cost with N-1 . The fault tolerance characteristics is highest because there is neither a centralized unit nor
a shared bus; as a negative figure of merit it should be pointed out that
increasing N the dedicated links are poorly utilized.

<u>Trees</u>

    Tree interconnection structures soffer some limitationsof ring net-
works (delay in message passing) and some limitationsof star architectures (bottlenecks at or near the root).
    A tree structure with B branches per node and $\bar{p}$ node levels from
the root to the leaves is shown in fig. 9.
    The total number of nodes is $N=(B^{\bar{p}}-1)/(B-1)$.
    This architecture is in general considered constituted by computing
nodes (leaves) and switching nodes (all other nodes of the tree). The
message delay is variable from 2 (message between to brother leaves) and
2 (p-1).
    From the point of view of fault-tolerance, the tree structure has
not a good figure of merit considering that a fail in a switching node
disconnects the whole subtree that has the failed node as root from the
rest of the structure and that subtree is divided in two subtrees not
communicating each other. The message traffic distribution in the tree
is not uniform and the root as well as the nodes near the root can become a bottleneck for the performances of the whole system.
    The analysis of tree architecturesfrom the point of view of cost
complexity and traffic density is not simple considered that the struc-

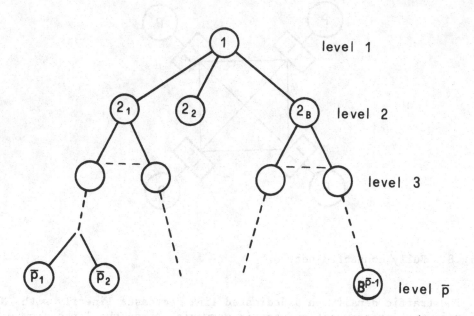

Fig.9. Tree connected system.

ture is not uniform, therefore the cost index derived in the next chapter
will not be applied to it.

4.2. Complex interconnection structures

The simple architectures discussed in the previous paragraph, except
perhaps the tree, can be used for small values of N .

For very large networks more complex interconnection topologies must
be used. These topologies have been investigated mainly from a theoretical point of view, but they will become realizable in the near future
when a complete node (CPU, memory and I/O interface) will be putted on
a single chip. The common characteristics of these complex structures
is that they are enbedded into a D-dimensional hypercube with the N
noded lieing on the W^D lattice points (W-wide, D-dimensional hypercube.

Nearest Neighbor Mesh (NNM)

This is a well known structure used in some realization of parallel machines and to implement parallel algorithms.

In this structure there are M nodes per dimension connected as in fig.10 for $D=2$, therefore $N=M^D$. Each node needs 2 switches per dimension hence the number of switches per node is 2D. In each dimension the distance between two nodes is variable between 1 and $M/2$ and the average distance is approximately $M/4$: the average delay for message passing in the M^D hypercube is hence $D \cdot M/4$.

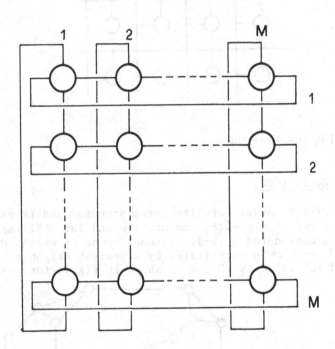

Fig.10.   Nearest neighbor mesh interconnection.

Spanning Bus Hypercube (SBH)

This structure is similar to the nearest neighbor mesh, with the difference that in each dimension the M nodes are connected to a bus of width $W=M$. Therefore $N=W^D$, the number of switches per node is D, the average delay in message passing is $\sim D-1$. In fig.11 a spanning bus hypercube with $D=2$ and $W=4$ is shown.

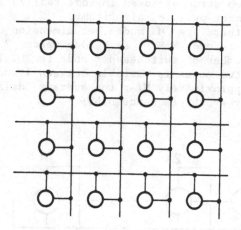

Fig.11.   Spanning bus hypercube.

Cube Connected Cycle (CCC)

This is one of the newest architectures proposed and is particular attractive for parallel algorithm computation and for VLSI implementation. The CCC can be considered a D-dimensional cube of width D where each of the 2^D vertices is substituted by a nearest neighbor mesh of dimension 1 and multiplicity D as shown in fig.12 for D=3 .

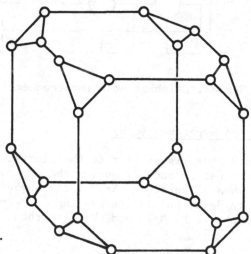

Fig.12.   Cube connected cycle.

Therefore $N = D \cdot 2^D$, each node is connected to three links, and the average delay in message passing is approximately 7D/4 .

D-Dimensional Array (DDA)

This is a new architecture proposed by myself, and is particular attractive to realize large distributed computers with relatively small cost. The DDA can be considered a SBH where each node is substituted by a global bus of width D . Therefore the total number of nodes is $N=D \cdot W^D$, the number of switches per node is 2, and the average delay in message passing is approximately $2(D-1)$.

The topology of a DDA with D=2 W=3 is shown in fig.13.

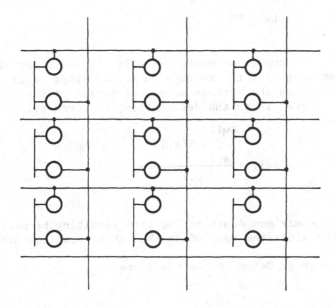

Fig.13.   DDA Architecture.

5. A MODEL FOR THE EVALUATION OF INTERCONNECTION STRUCTURES

5.1. Key factors

Let be given N processors (nodes) to be connected one another by means of L links or busses. $K(i)$ is the number of poles of the switch attached to processor P_i for connecting P_i to $K(i)$ links or busses; in the most general case $K(i)$ is function of i , whereas in uniform and symmetric structures (USS) K is constant.

For bus connected structures let $W(j)$ be the width of bus j , defined as the number of nodes connected to the bus. The link connected structures can be considered a particular case of bus connected structures (W=2) for most of the considerations developed in the sequel.

Unless explicity specified, the results presented hold for both type of structures.

It is easy to verify that the following general relation hold:

$$\sum_{j=1}^{L} W(j) = \sum_{i=1}^{N} K(i) \tag{1}$$

In particular if $K(i)=K$ and $W(j)=W$, as it is true for USS , the (1) becomes:

$$LW = NK \tag{1'}$$

Let $R(s,i)$ be the number of nodes rechable in s steps with a shortest path from node i : the average number of steps $S(i)$ in message routing, defined as the average number of busses utilized by a message from the source i to the destination, is given by:

$$S(i) = \frac{\sum_{s=1}^{s_M(i)} s \cdot R(s,i)}{N-1} \tag{2}$$

where $s_M(i)$ is the maximum depth of the tree resulting by means of a breath-first-search algorithm applied to the interconnection graph rooted in node i .

The structure average delay S is therefore:

$$S = \frac{\sum_{i=1}^{N} s(i)}{N} \tag{3}$$

For USS $R(s,i)$ is independent from i , hence:

$$S = \frac{\sum_{s=1}^{s_M} sR(s)}{N-1} \tag{3'}$$

S is a crucial parameter for the performance of the overall structure. In fact S influences the average number of messages per unite time (average traffic density) on busses and nodes as well as the corresponding average delay in message passing.

The considerations in the sequel are developed under the following assumptions:

- the traffic is uniformely distributed, i.e. for every node pair the communication probability is the same

- the traffic generated by every node is the same and is considered u-nitary.

Let $T_B(i,h)$ be the traffic density for message passing on bus $h(h=1,\ldots,K(i))$ connected to node i[1] ; the average traffic density (overhead traffic) on node i is:

$$T_N(i) = \sum_{h \in \beta(i)} \frac{T_B(i,h)}{W(h)} - 1 \qquad (4)$$

where $\beta(i)$ is the set of busses connected to node $i (|\beta(i)|=K(i))$. For USS $T_B(i,h)=T_B=NS/L$, hence:

$$T_B = \frac{WS}{K} \qquad (5)$$

$$T_N = \frac{K}{W} T_B - 1 = S - 1$$

It is worthwhile noticing that T_B and T_N depends through S on the network topology and size. Moreover the ratio between T_B and T_N+1 (global traffic on each node) is:

$$\frac{T_B}{T_N+1} = \frac{N}{L} \qquad (6)$$

i.e. for structures with $N>L$ (typically bus connected structures) the bottlenecks are the busses.

The average delay time Δ_B at a node for accessing a bus, under standard conditions (Poisson arrivals at each node, indipendence as-sumptions) is:

$$\Delta_B = \frac{1}{T_M-T_B} \qquad \text{where } T_M = \text{bus capacity expressed as number of messages per unit time.}$$

---

(1) For sake of simplicity we have used the notation $T_B(i,h)$ to i-dentify explicitely the bus; this does not mean that T_B depends only on i and h .

The average end-to-end delay Δ is then:

$$\Delta = S\Delta_B \tag{7}$$

In the analysis of asymptotic behavior of interconnection structures, i.e. increasing number of nodes, for almost all the structures of practical interest, T_B is a monothonic increasing function of N. Therefore it is meaningless to consider T_M constant; a resonable assumption consists on supposing constant the average delay time on a bus: $\Delta_B = \alpha_1$ [2]. With this assumption the end-to-end delay becomes:

$$\Delta = \alpha_1 S \tag{8}$$

and the bus capacity:

$$T_M = T_B + \frac{1}{\alpha_1} \tag{9}$$

5.2. The cost index

A cost index suitable for comparing different architectures will be defined as the product of two terms:

I = hardware cost per node x message average time delay.

The smaller is the index I, the better is the tradeoff between structural complexity of the network and its performance.
The average time delay is given by the (8). The hardware cost per node, in the general case, can be considered the sum of four cost terms:

i)   node-bus interface channels: C_c

ii)   busses: C_B

iii) nodes : C_N

iv)   bus arbiters: C_A

---

[2]   In the sequel all constants will be indicated by α_i (i=1,2,...).

C_c is assumed to be proportional to the number of channel interfaces time the width of the busses:

$$C_c = \alpha_2 \sum_{i=1}^{N} \sum_{j \in \beta(i)} W(j)$$ (10)

In fact, for each interface channel the transmission speed must be proportional to the number of nodes connected to the same bus.
The cost of a bus j is assumed to be proportional to its capacity $T_M(j)$ hence:

$$C_B = \alpha'_3 \sum_{j=1}^{L} T_M(j) = \alpha_3 \sum_{j=1}^{L} T_B(j) + \alpha_4 L$$ (11)

With similar considerations we obtain:

$$C_N = \alpha_5 \sum_{i=1}^{N} (T_N(i) + 1)$$ (12)

$$C_A = \alpha_6 \sum_{j=1}^{L} W(j)$$ (13)

For USS the (10), (11), (12) and (13) simplify in:

$$C_c = \alpha_2 NKW$$ (10')

$$C_B = \alpha_3 T_B L + \alpha_4 L$$ (11')

$$C_N = \alpha_5 NS$$ (12')

$$C_A = \alpha_6 LW$$ (13')

Finally the index I results:

$$I = \frac{\Delta}{N} (C_c + C_B + C_N + C_A)$$ (14)

and for USS:

$$I = \alpha_1 S(\alpha_2 KW + \alpha_3 S + \alpha_4 \frac{K}{4W} + \alpha_5 S + \alpha_6 K)$$ (14')

Asymptotically we obtain:

$$I = O(S(KW+S)) \tag{15}$$

5.3. Lower bounds for the cost index of USS

In the previous paragraph we have obtained a cost index depending only on K,W,S. In order to evaluate different architectures it can be useful to compare the cost index behavior against a theoretical lower bound. It is common, in order to simplify the formulae, to consider the asymptotical behavior of the cost index as function of N. Therefore we must express W,K and S as function of N, and this will be done for the specific architectures analysed in the next section. As far as the lower bound is concerned, we can derive a lower bound for I starting from a lower bound for S, as function of N, W and K.

As stated by (15), the cost index depends on S and $K \cdot W$. Now, if $K=1$ the only interconnecting structure is the single bus, and since for this structure $W=N$ and $S=1$, the cost index becomes $I=O(N)$. If $K=2$ and $W=2$, the only interconnecting structure is the linear array, for which $S=N/4$ and $I=O(N^2)$. On the other hand, if $S=1$ the interconnecting structure must have either $W=N$ and $K>1$ or $W=2$ and $K=N-1$; in both cases $I=O(N)$.

Therefore in order to obtain a cost index better than $O(N)$ it is necessary that $S>1$, $K \geq 2$, $W \geq 2$, and $K \cdot W \geq 6$.

Theorem 1: $S = \Omega \left(\dfrac{\log N}{\log W + \log K} \right)$

Proof. consider the maximum number of nodes reachable in p steps for given values of K and W; $R(p)$ is bounded by:

$$R(p) \leq K(W-1) \left[(K-1)(W-1) \right]^{p-1} \tag{16}$$

Let p_{max} be the maximum number of steps needed to reach all the nodes if the strict inequality in the (16) holds for at least a p, and $\bar{p}$ the same if the (16) holds with the equality for all p except possibly $\bar{p}$: $p_{max} \geq \bar{p}$.

Since $S = \sum\limits_{p=1}^{p_{max}} p R(p)/(N-1)$, substituting the right-hand side of the

(16) to $R(p)$ this implies an increase of the number of nodes multiplied for smaller p , hence

$$S > \frac{K(W-1)}{N-1} \sum_{p=1}^{\bar{p}} p\gamma^{p-1} \quad \text{where} \quad \gamma=(K-1)(W-1)$$

Since $\quad N-1 \leq \sum\limits_{p=1}^{\bar{p}} K(W-1)\gamma^{p-1} \qquad$ we can state:

$$S > \frac{\sum\limits_{p=1}^{\bar{p}} p\gamma^{p-1}}{\sum\limits_{p=1}^{\bar{p}} \gamma^{p-1}} = \frac{\frac{d}{d\gamma}\left|\gamma\sum\limits_{p=o}^{\bar{p}-1}\gamma^{p}\right|}{\sum\limits_{p=o}^{\bar{p}-1}\gamma^{p}} = \bar{p}\,(\frac{\gamma^{\bar{p}}}{\gamma^{\bar{p}}-1} - \frac{1}{\bar{p}(\gamma-1)}) \geq c\bar{p}$$

where $\ 1/2 \leq c \leq 1 \ $ and $\ c \to 1 \ $ for increasing values of $\ \bar{p} \ $ and/or $\ \gamma \ $.
On the other hand, with the hypothesis on $\ K \ $ and $\ W \ (KW \geq 6)$, $\ N \leq 4\gamma^{\bar{p}} \ $, hence:

$$\bar{p} \geq \frac{\log N - 2\log 2}{\log \gamma} \qquad \text{and} \qquad S \geq \frac{\log N}{\log \gamma}$$

Substituting in the (15) the expression $\log N/\log\gamma$ instead of S, we obtain a lower bound for I as function of N and γ :

$$I = \Omega\,((\frac{\log N}{\log \gamma})^2 + \gamma\,\frac{\log N}{\log \gamma}) \tag{15'}$$

For $\ 2 \leq \gamma \leq N \ $ the two components of the right-hand side of the (15') vary as shown in fig.14.
Remark that $\gamma = $ constant implies:

$$I = \Omega\,(\log^2 N)$$

For $\gamma = f(N)$ we can derive a different lower bound.
In fact for $\gamma < \bar{\gamma}$ the first component of the (15') dominates the second one, whereas for $\gamma > \bar{\gamma}$ the second one is dominant. For $\gamma = \bar{\gamma}$ the two components are equal, therefore $\bar{\gamma}$ minimize the order of the lower bound of I. On the other hand $\bar{\gamma}$ depends on N as $\bar{\gamma}^{\bar{\gamma}} = N$, and since

$$0.47\,\frac{\log N}{\log\log N} < \bar{\gamma} < 1.37\,\frac{\log N}{\log\log N} \ , \quad \bar{\gamma} = \theta\,(\frac{\log N}{\log\log N})$$

Therefore we have proved the following:

Theorem 2: $\ I = \Omega\,((\frac{\log N}{\log\log N})^2) \quad .$

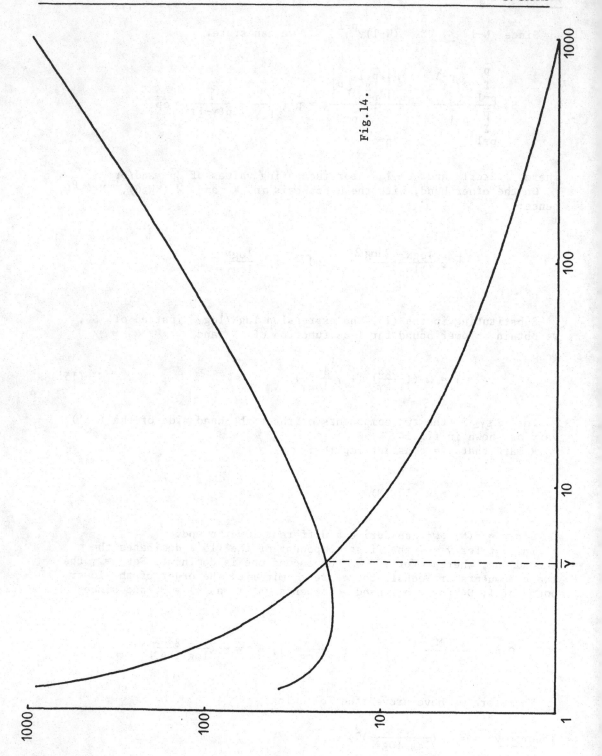

Fig.14.

6. COMPARISON OF THE ARCHITECTURES

With the cost index derived in the previous section, it is possible to compare the asymptothic behavior of the architectures introduced in section 4 with the exception of star and tree since these architectures are not USS.

The parameters of the architectures presented are:

1) Global bus : $K = L$; $W = N$; $S = 1$

2) Ring : $K = 2$; $W = 2$; $S = N/2$

3) Fully connected: $K = N-1$; $W = 2$; $S = 1$

4) NNM : $K = 2D$; $W = 2$; $N = M^D$; $S = DM/4$

5) SBH : $K = D$; $N = W^D$; $S \simeq D-1$

6) CCC : $K = 3$; $W = 2$; $N = D \cdot 2^D$; $S \simeq 7D/4$

7) DDA : $K = 2$; $N = D \cdot W^D$; $S \simeq 2(D-1)$

The cost index has been applied to the architectures 1÷7 with the assumption $M=W=D$, to simplify the formulae.

The behavior of the cost indexes are:

$$I_1 = O(L \times N); \quad I_2 = O(N^2); \quad I_3 = O(N)$$

$$I_4 = O((\frac{\log N}{\log \log N})^4); \quad I_5 = O((\frac{\log N}{\log \log N})^3)$$

$$I_6 = O(\log^2 N); \quad I_7 = O((\frac{\log N}{\log \log N})^2)$$

From the above cost indexes we can confirm that the USS architectures lieing on the lattice points within a hypercube have a better behavior than the simpler ones.

The DDA reaches the theoretical USS lower bound for bus interconnected structures, and the CCC reaches the lower bound for link interconnected structures.

Note that the DDA as all busses interconnected structures has a limitation on the width of the spanning and local busses and therefore on D. However small values of D produce vast number of nodes. For example D=5 implies N=15´625, D=10 implies N=10^11

Moreover W_s, the width of the spanning busses, can be different from D. This fact in practice is very useful to vary in a finer way the number of nodes.

7. MESSAGE PASSING ORGANIZATION

Loosely coupled multiprocessor systems found the cooperation between processes allocated into distinct nodes on the explicit transmission of a message from the source process to the destination one (3).

This approach to the cooperation permits to achieve a more reliable system since there are two processes, source and destination, that co-operate in an explicit way in the information exchange. This fact allows a more general and rigorous control on the information passing since the logical channel assigned to the two partner processes can be organized with special features (type of data, type of synchronization and so on) that increase the robusteness and the fault-tolerance aptitude of the whole system.

To be more specific the communication mechanisms can be classified according to some features:

a)   synchronous communication: in this form of communication a "rendez-vous" is established between the source and the destination pro-cesses, i.e. the message exchange occurs if and only if both part-ners are ready to perform it;

b)   asynchronous communication: in this form of communication the source process puts the message into a message buffer of the destination process without waiting for any action from the latter.

In the synchronous communication the "rendez-vous" can be <u>limited</u> to the message transfer phase (CSP-like languages) or <u>extended</u> when the source process performs, by means of the message, a remote procedure-call and then waits for the results (ADA and DP-like languages).

The asynchronous form of communication is more efficient, considered that the source process must not synchronize itself with the destination one, whereas the synchronous form is more safe, considered that some recovery action can be immediately undertake in case of logical fail-ures.

There are other features that can be introduced for characterizing the kind of communication as the characteristic of the logical channel between the partner processes and the type of data exchanged via the channel. From the point of view of system performance loosely coupled systems lack in efficiency of cooperation since both sender processor and receiver processor participate to the message passing action, the former with an output routine, the latter with an input one.

In this aspect we can consider tightly coupled systems as ones in which the cooperation is performed via processor-memory transfers, whereas loosely coupled systems as ones in which the cooperation is performed via processor-processor transfers.

A better solution to this problem consists in providing each pro-cessor with DMA channels or specialized I/O processors which are dedicated to the message passing activity. This approach permits a much faster transmission, that can be considered a memory-memory trans-fer. Moreover this activity, as it is transparent to the processors can be concurrent with the process running.

A possible organization of memory-memory transmission is shown in fig.15, for a bus connected architecture.

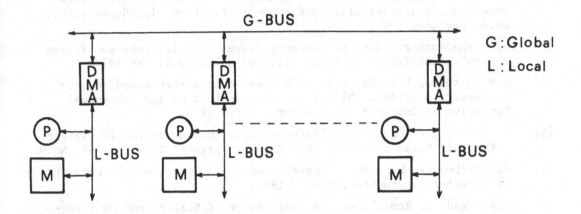

Fig.15. Distributed system organization.

Note that during the transmission the partner processors are slowed down   due to the bus stealing operated by the DMA channels. Better performances can be achieved if the memory area dedicated to I/O message buffers is a dual port memory block as shown in fig.16. In this case the DMA channel becomes a specialized I/O processor. The slowing down of partner processors is dratically reduced since the normal activity of them is carried out on the private memory.

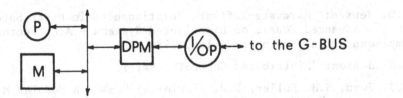

Fig.16. Improved organization of distributed systems.

References

|1| D.P. Agrawal et alii: "A survey of Communication Processor Systems".
     Proceedings of COMPSAC/78.

|2| G.A. Anderson, E.D. Jensen: "Computer Interconnection Structures:
     taxonomy, characteristics and examples". ACM Computing Surveys, Vol. 7,
     No. 4, December 1975.

|3| D.P. Bhandarkar: "Some performance issues in multiprocessor system
     design". IEEE Trans. on Computers, Vol.C-26, No.5, May 1977.

|4| A.M. Despain, D.A. Patterson: "X-tree: a tree structured multi-
     processor computer architecture". Proceedings of the 5th Annual
     Symposium on Computer Architecture, April 1978.

|5| N. Jovic, G.W. Conturier: "Interprocessor Communication in systems
     with distributed control". IEEE Proc., September 1977, Vol.65, No.9.

|6| G.J. Lipovsky, K.L. Doty: "Developments and Directions in Computer
     Architecture". Computer, August 1978.

|7| C.A. Mead, M. Rem: "Cost and performance of VLSI computing struc-
     tures". IEEE Proc., Vol.SC/4, No.4, April 1979.

|8| L.D. Wittie: "A distributed operating system for a reconfigurable
     network computer". Proceedings of the 1st International Conference
     on Distributed Computing Systems, Huntsville, Alabama, October
     1979.

|9| G. Cioffi, P. Corsini, G. Frosini, L. Lopriore: "MuTEAM: Architec-
     tural Insights of a Distributed Multimicroprocessor System". Proc.
     of 11th Fault-Tolerant Computing Symposium, June 1981.

|10| F. Baiardi, A. Fantechi, A. Tomasi, M. Vanneschi: "Mechanisms for
     a Robust Distributed Environment in the MuTEAM Kernel". Proc. of
     11th Fault-Tolerant Computing Symposium, June 1981.

|11| P. Ciompi, F. Grandoni, L. Simoncini: "Distributed Diagnosis in
     Multimicroprocessor Systems: The MuTEAM Approach". Proc. of 11th
     Fault-Tolerant Computing Symposium, June 1981.

|12| P. Denning: "Fault Tolerant Operating Systems". ACM Computing Sur-
     veys, 8, 4, December 1976.

|13| E.D. Jensen: "Hardware-Software Relationships in Distributed Sys-
     tems". Advanced Course on Distributed Systems - Architecture and
     Implementation, Springer-Verlag, 1980.

|14| E.D. Jensen: "Distributed Control". Ref.1.

|15| R.J. Swan, S.H. Fuller, D.P. Siewiorek: "Cm* - A Modular Multimi-
     croprocessor". Proc. AFIPS 1977, NCC, 46.

|16| C.A. Monson, P.R. Monson, C.P. Marshall: "A Cooperative Highly
     Available Multiprocessor Architecture". Proc. COMPCON 79 Fall,
     September 1979.

|17| R.E. Bryant, J.B. Dennis: "Concurrent Programming". MIT Report,
      MIT Cambridge Mass., 1979.

|18| D.L. Russel: "State Restoration in Systems of Communicating Pro-
      cesses". IEEE Trans. Software Eng., Vol.SE-6, 2, March 1980.

|19| D.L. Parnas: "On the Criteria· to be Used in Decomposing Systems
      into Modules". Comm. of the ACM, 15, 12, December 1972.

|20| C.A.R. Hoare: "Communicating Sequential Processes". Comm. of the
      ACM, 21, 8, August 1978.

|21| C.A.R. Hoare: "A Model for Communicating Sequential Processes".
      Oxford University Report, July 1979.

A VLSI SORTER

C.K. Wong

IBM Thomas J. Watson Research Center,
P.O. Box 218, Yorktown Heights, NY 10598

1. Introduction

Sorting is one of the most important operations in data processing. It is estimated that in data processing centers, over 25 percent of CPU time is devoted to sorting [6]. Many sequential and parallel sorting algorithms have been proposed and studied [1-3, 5, 6, 8, 10, 13-18]. Implementation of various sorting algorithms in different hardware structures has also been investigated [2-4, 7, 8, 11, 12, 16, 18].

In this paper, we describe a sorter where the sorting time is completely overlapped with the input/output time. It has complete parallel operations and process data in a pipelined fashion. It can sort in both ascending and descending order and can overlap the sorting time of two consecutive input sequences. Because of the regularity of its structure, it is most suitable for VLSI implementation. A detailed implementation is presented to illustrate the basic principle. Further optimization in various aspects of the design is clearly possible.

2. Principle

The sorter consists basically of a linear array of n/2 cells. (we assume n is even), each of which can store two items of the sequence to be sorted (Figure 1). The initial sequence is input to the sorter one item at each step. After the input of the last item, the data flow direction is reversed and the sorted sequence is then output, also serially. Each step is executed synchronously and simultaneously by all the cells and has two phases:

1) Compare: the two items in each and every cell are
compared to each other,

2) Transfer: subject to the result of the comparison,
                 desired sorting order (ascending or descending)
                 and the sorting state (input or output), either
                 one of the two items is transferred to the neighbor
                 cell and receives one from the other neighbor cell.

The sorter not only processes the items of a given sequence in a pipelined fashion, but also sorts different sequences in a pipelined way, i.e., while one sorted sequence is being output, a new sequence could be input at the same time from the other end of the sorter. This way, the I/O time of the sequence is completely absorbed by the sorting time needed by another.

· Figure 2 is an example of the sorting of a sequence in ascending order. "∞" represents the largest item possible. At the input stage the larger of the two items in each cell is transferred down; while at the output stage the smaller of the two is transferred up. Note that at the end of the input stage (step 6), the smallest item must be in the top cell, the second smallest must be in either the top or the second cell. In general, the i-th smallest item must be in one of the top i cells. This is why the output sequence is sorted.

The same principle applies to descending sort; we have only to replace "∞" by "-∞", the smallest item and interchange larger and smaller. (It will be shown later that it is not necessary to flood the sorter initially with either "∞" or "-∞". (See Figure 14.))

Let A,B be the two items stored in a cell. Let $M=Max(A,B)$, $m=min(A,B)$. If we consider the sorting of an isolated sequence, and the sequence is input and output through the top (top sequence), the specific action in the transfer phase can be summarized as follows,

Sort order \ Stage	Input (Down)	Output (Up)
Ascending	M moves down to next cell (M↓)	m moves up to next cell (m↑)
Descending	m moves down to next cell (m↓)	M moves up to next cell (M↑)

Table 1

If the sequence is input and output through the bottom port of the sorter (bottom sequence), the table would be:

Sort Order \ Stage	Input (Up)	Output (Down)
Ascending	M↑	m↓
Descending	m↑	M↓

Table 2

A fact to be noted is that the roles of M and m are interchanged when we consider a descending as opposed to ascending sort.

When we overlap the output of a sequence with the input of another, it is clear from Tables 1 and 2 that the transfer actions are different for the two sequences. For example, for an ascending sort, in the upward movement, we have m↑ for the output (top) sequence and M↑ for the input (bottom) sequence.

For this distinction, we attach a flag to each item when it is input: "0" ("1") to items in top (bottom) sequence. This flag will be considered part of the item, in the comparison as well as in the transfer. And we obtain the table on transfer actions as follows:

Data Movement \ Tag bits	0 0	1 1	0 1
Downward	M↓ (m↓)	m↓ (M↓)	M↓ (M↓)
Upward	m↑ (M↑)	M↑ (m↑)	m↑ (m↑)

Table 3

The parenthesized entries correspond to descending sort. The third column represents the frontier cell between the two sequences. If we include the tag bit as the most significant bit of the items for the purpose of comparison, the item from a bottom sequence with tag bit = 1 will be always M and the two sequences will always be kept separate. An example of the sorting with the added tag bits is shown in Figure 3.

3. Logic Design

Throughout this paper, the cell array of the sorter will be represented vertically. Each cell, containing two w-bit items, is a horizontal linear array (row) of w dibit-cells. The overall topological layout is shown in Figure 4. In actual physical layout, a carpenter folding [9] of the cell array might be needed to obtain a more square-shaped chip.

Dibit-cell. Each such cell is a compare/steer unit for two bits, one from each of the two items A and B, representing the same bit position. Figure 5 is the block diagram of a dibit-cell. In downward (upward) movements, after comparison, one of the two bits will be shifted out on line a (b) to the next (previous) cell, while a bit from the previous (next) cell is being shifted in on line I (O). In that Figure, the terms "input" and "output" refer to a top sequence, and the controls are indicated for an ascending sort.

A circuit schematic of a dibit-cell is shown in Figure 6. The precharged carry-propagate-type comparator is shown together with the two bit-cells. It should be noted that every bit-cell of item A (B) in a cell row is controlled by the same 4 signals C_1, C_2, C_3 and C_4 (C_1', C_2', C_3' and C_4'), so that all the bits of an item are recycled or shifted at the same time.

The comparators of the dibit-cells in a cell row are chained as in Figure 7. C is the comparison result of items A and B, i.e., C=1 if item A$\geq$ item B, C=0 otherwise. The comparison carry chain is precharged during clock phase ϕ_1 (gates W and Y in Figure 6).

Control. To illustrate, let us consider an ascending sort with a top sequence. Each cell is a 2-inverter loop controlled by 4 gates using a 2-nonoverlapping-phase clock. The required gatings for different situations with A$\geq$B (i.e. comparison result C=1) are shown in Figure 8. In the case of A<B, just interchange the gatings for A and B. The boolean expressions obtained are listed as follows,

$$C_1 = \phi_2 Ia + \phi_1 \bar{I}\bar{a} \qquad C_1' = \phi_2 I\bar{a} + \phi_1 \bar{I}a$$

$$C_2 = \phi_2 I\bar{a} + \bar{I} \qquad C_2' = \phi_2 Ia + \bar{I}$$

$$C_3 = I + \phi_2 \bar{I}a \qquad C_3' = I + \phi_2 \bar{I}\bar{a}$$

$$C_4 = \phi_1 Ia + \phi_2 \bar{I}\bar{a} \qquad C_4' = \phi_1 I\bar{a} + \phi_2 \bar{I}a$$

I=1(0) indicates the downward (upward) movement. a is the boolean variable which takes opposite values (0 and 1) in opposite situations:

-   ascending (Opt=0) versus descending sort (Opt=1),

-   top (SR=0) versus bottom sequence (SR=1),

-   and A$\geq$B (comparison carry C=1) versus A<B (C=0).

It follows that a is the exclusive-OR of C, SR and Opt, i.e.

Opt \ SR	0	1
0	C	$\overline{C}$
1	$\overline{C}$	C

See Figure 9 for the circuit schematic of the cell control.

To have homogeneous and regular cells, we have avoided the explicit use of the tag bit combination to distinguish top and bottom sequences (Table 3), instead we have a bidirectional double shift-register chain, whose contents move up and down in synchrony with those of the cells and whose output at each level is taken to be SR, as shown in Figure 10, so that an item of a top (bottom) sequence is always chaperoned by SR=0 (1). A slight complication occurs at the frontier. The desired transfer action table is then

	Ascending			Descending		
Tag-bits	00	11	01	00	11	01
Down	M↓	m↓	M↓	m↓	M↓	M↓
Up	m↑	M↑	m↑	M↑	m↑	m↑
SR	0	1	0	0	1	1

Table 4

The reader could easily check out from Figure 10 that the two extra unidirectional shift registers at the two ends are needed to fulfill the requirement of the third column in both ascending and descending sort.

4. Timing

We use a 3-nonoverlapping-phase clock as shown in Figure 11. During phase ϕ_1, the transfer bit is read out from cell(i) while the other bit is recycled and the comparison carry chain precharged (Figure 12). During ϕ_2, the transfer bit is written into the next cell (i+1 or i−1) while the other bit is making a full recycle and the comparison taking place. At ϕ_3, the comparison result signal is fed into the control circuit of each cell.

In addition, phase ϕ_3 is needed (see Figure 13)

(1) for the transition from up to down and down to up stages,

(2) for the initialization,

(3) and to avoid racing condition in the loop of comparator, control, and bit cell.

5. Initialization

Before the beginning of a sort, instead of initializing all the cells with "∞" or "-∞", it is necessary only to fill in the two border cells with tags distinct from the tags of the sequence coming in, together with appropriate setting of the comparison shift registers as in Figure 14. Recall that top (bottom) sequences have tag bit "0" ("1"). So here "∞" ("-∞") represents any number with tag bit "1" ("0"). It could be easily checked from Table 4 and e.g. Figure 14e that these initializations are indeed adequate.

All the initial values are injected into the sorter during clock phase ϕ_3.

6. Concluding Remarks

1) The circuits are drawn up as if the wires connecting dibit-cells of rows i and i+1 have enough capacitance to store the transfer bit. If they do not, it would be a simple matter to add to them connection inverters. Without the inverters, comparisons on adjacent row cells must be implemented differently. Indeed, as can be seen in Figure 6, a bit leaving a cell is in complemented form than when it was input. Therefore, to produce the same comparison carry output we need to invert the roles of A and $\overline{A}$, and also B and $\overline{B}$ as in Figure 15. A redrawn global block diagram is shown in Figure 16 where the alternation between adjacent rows is

clearly indicated. Note also that an even number of rows is recommended so that data are input and output in "true" form. (Otherwise either the top or bottom would be in "false", i.e. negated form.)

2) For our implementation (Figure 6) we have a device count of 26 for a dibit-cell, i.e. 13 per bit versus 6 in today's 16K static RAM. So a sorter chip would have very likely a capacity up to 8K bit or 256 32-bit cells. The sorter can be trivially extended to handle key/pointer pair by simply omitting the compare logic on the portion of the storage cell associated with the pointer. (Then it will require only 8 devices per pointer bit.)

3) We can use the sorter to merge two sorted strings by repeatedly passing them through the sorter in an appropriate way. For example, the generalized odd-even merge algorithm described in ([6], p.241, Excerise 38) can be employed for this purpose.

References

[1] K.E. Batcher, "Sorting Networks and their Applications," AFIPS Conference Proc., Vol.32, 1968, Spring Joint Computer Conference, pp.307-314, Apr. 1968.

[2] T.C. Chen, V.Y. Lum, and C. Tung, "The Rebound Sorter: An Efficient Sort Engine for Large Files," Proc. 4th VLDB, pp.312-318, Sept. 1978.

[3] K. Chung, F. Luccio, and C.K. Wong, "On the Complexity of Sorting in Magnetic Bubble Memory Systems," IEEE Trans. Comput., Vol.C-29, No.7, pp.553-563, July 1980.

[4] M.J. Foster and H.T. Kung, "The Design of Special-Purpose VLSI Chips," IEEE Computer, Vol.13, No.1, pp.26-40, Jan. 1980.

[5] D.S. Hirschberg, "Fast Parallel Sorting Algorithms," Communications of the ACM, Vol. 21, No.8, pp.657-661, Aug. 1978.

[6] D.E. Knuth, "The Art of Computer Programming," Vol.3, 'Sorting and Searching', Reading, Massachusetts, Addison-Wesley, 1973.

[7] H.T. Kung, "The Structure of Parallel Algorithms," Advances in Computers, Vol.19, Academic Press, pp.65-112, 1980.

[8] D.T. Lee, H. Chang, and C.K. Wong, "An On-Chip Compare/Steer Bubble Sorter," IEEE Trans. Comput., Vol.C-30, No.6, pp.396-405, June 1981.

[9] C.E. Leiserson, "Area-effficicent graph layouts (for VLSI)," Proc. 21st Annual Sym. on Foundations of Computer Science, IEEE, Oct. 13-15, 1980.

[10] H. Lorin, "Sorting and Sort System," Reading, Massachusetts, Addison-Wesley, 1975.

[11] C. Mead and L. Conway, "Introduction to VLSI Systems," Reading, Massachusetts, Addison-Wesley, 1980.

[12] A. Mukhopadhyay, "Hardware Algorithms for Nonnumeric' Computation," IEEE Trans. Comput., Vol.C-28, No.6, pp.384-394, June 1979.

[13] D.E. Muller and F.P. Preparata, "Bounds to Complexities of Networks for Sorting and for Switching," JACM, Vol.22, No.2, pp.195-201, Apr. 1975.

[14] D. Nassimi and S. Sahni, "Bitonic Sort on a Mesh-Connected Parallel Computer," IEEE Trans. Comput., Vol.C-28, No.1, pp.2-7, January 1979.

[15] F.P. Preparata, "New Parallel-Sorting Schemes," IEEE Trans. Comput. Vol.C-27, No.7, pp.669-673, July 1978.

[16] Y. Tanaka, Y. Nozaka, and A. Masuyama, "Pipelined Searching and Sorting Modules as Components of a Data Flow Database Computer," Proc. IFIP '80, pp.427-432, October 1980.

[17] C.D. Thompson and H.T. Kung, "Sorting on a Mesh-Connected Parallel Computer," Communications of the ACM, Vol.20, No.4, April 1977.

[18] H. Yasuura, N. Takagi, and S. Najima, "The Parallel Enumeration Sorting Scheme for VLSI," IEEE Trans. Comput. (to appear).

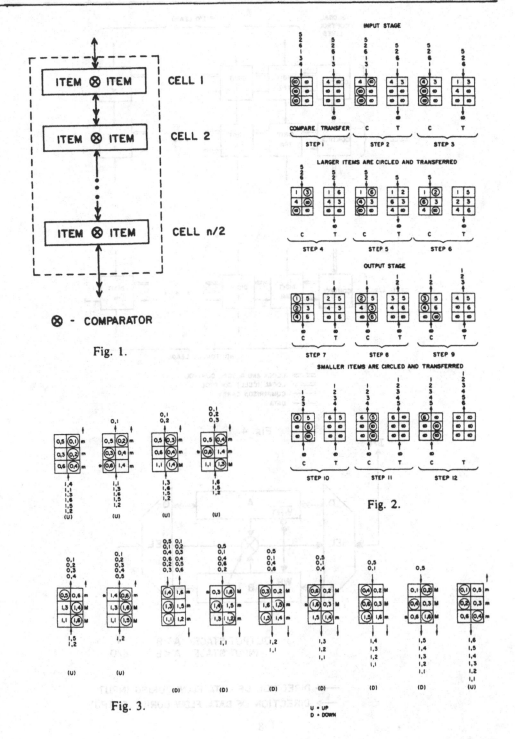

Fig. 1.

Fig. 2.

Fig. 3.

U = UP
D = DOWN

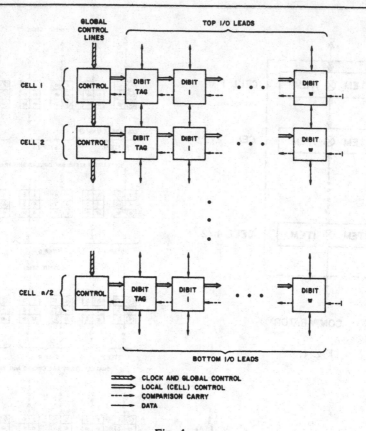

Fig. 4.

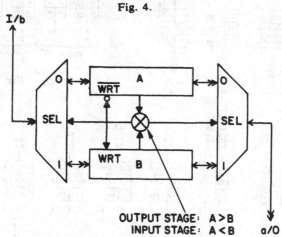

OUTPUT STAGE: A > B
INPUT STAGE: A < B

⟶ DIRECTION OF DATA FLOW DURING INPUT
⟶ DIRECTION OF DATA FLOW DURING OUTPUT

Fig. 5.

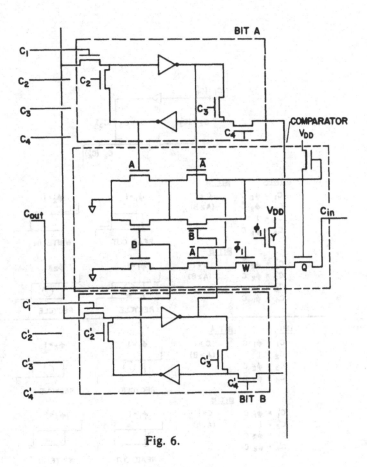

Fig. 6.

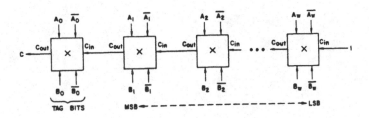

Fig. 7.

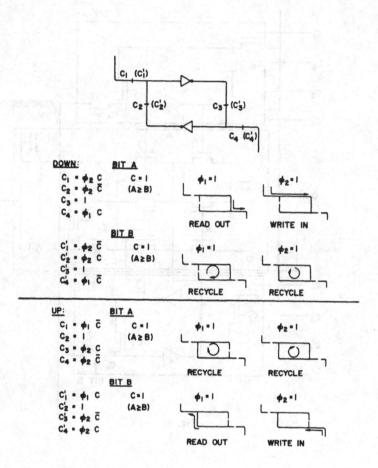

Fig. 8.

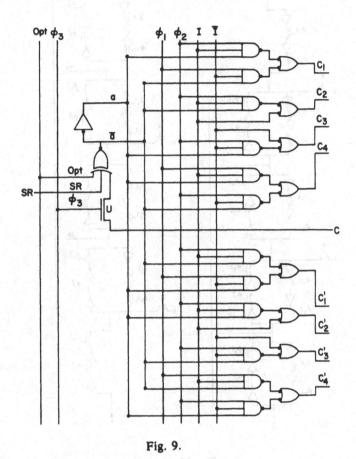

Fig. 9.

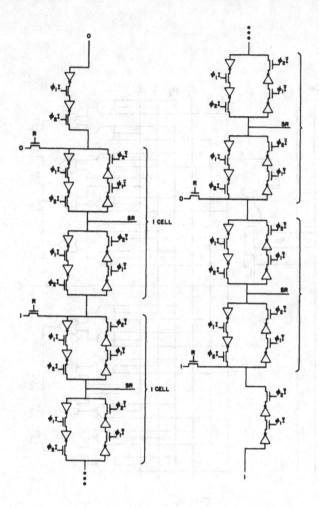

Fig. 10.

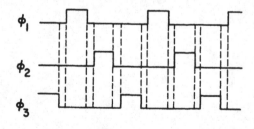

Fig. 11.

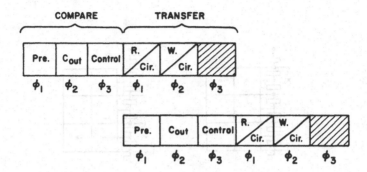

Pre.   = Precharging the comparison carry line.

C_{out} = Compute C_{out} in the comparators for bit pairs, obtaining C.

Control = C is fed into the control circuit of C_1, C_2, C_3, C_4, and C_1', C_2', C_3', C_4'.

R.   = Read the transfer bit out to the next cell (down or up).

W.   = Write in the transfer bit from the other next cell (up or down).

Cir.   = The stay bit is recycled in the cell.

Fig. 12.

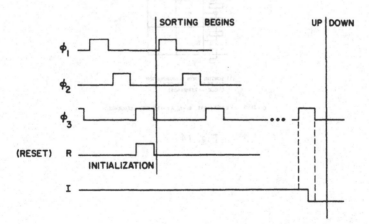

Fig. 13.

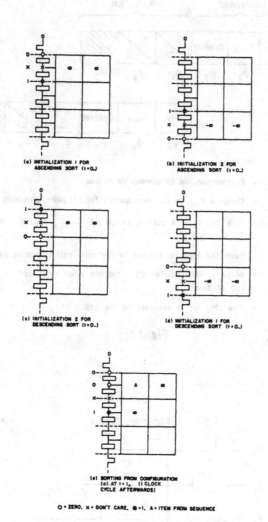

(a) INITIALIZATION 1 FOR
ASCENDING SORT (t = 0...)

(b) INITIALIZATION 2 FOR
ASCENDING SORT (t = 0...)

(c) INITIALIZATION 2 FOR
DESCENDING SORT (t = 0...)

(d) INITIALIZATION 1 FOR
DESCENDING SORT (t = 0...)

(e) SORTING FROM CONFIGURATION
(a) AT t = t₀.   (1 CLOCK
CYCLE AFTERWARDS)

O = ZERO, X = DON'T CARE, ⊕ = 1, A = ITEM FROM SEQUENCE

Fig. 14.

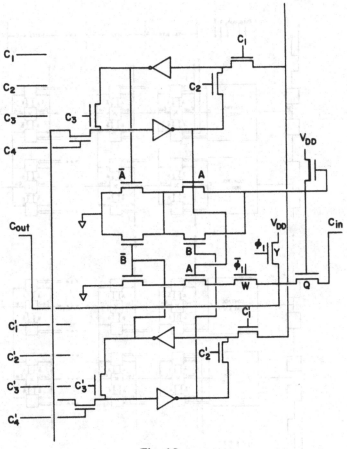

Fig. 15.

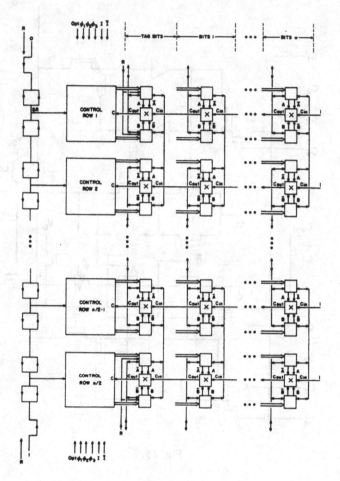

Fig. 16.

TOWARDS A THEORY OF VLSI LAYOUT
A Selected Annotated Bibliography

F.P. Preparata
University of Illinois at Urbana-Champaign

As the complexity of digital systems grew, the need arose for efficient packaging of the system components. Starting in the mid 50's, individual devices were placed on printed-circuit boards. As the size and the complexity of the modules grew - from individual devices to integrated circuits - the princed-circuit board has preserved to this day its function as a fundamental packaging level in the assembly of digital systems. However, a new layout horizon emerged, represented by the internal structure of the modules themselves (chip complexity). Today, the latter is the predominant problem in system layout: however, in spite of largely different feature sizes, the two environments - VLSI chip and printed-circuit board - are both governed by analogous sets of rules for the layout of wires on a regular grid. The highly structured layout medium and the necessity to cope with problems of increasing size motivated the development of automatic techniques and stimulated the attending research.

In general, the layout problem consists of two major subproblems: the *placement* of modules and the *routing* of wires to interconnect these modules. The identification of modules represents a hierarchical approach to layout: once the internal layout of the modules has been defined, the modules become unalterable geometric shapes, and only their external interconnection remains to be desired. Considerable flexibility exists as to the choice of modules: a particularly successful approach consists of adopting modules of identical shape and size and to place them on a regular grid (gate-array approach).

Although many approaches view placement and routing as two consecutive activities (see, e.g. reference [22] below), some gate-array-
oriented methods attempt the simultaneous solution of the two problems.
The most recent work along these lines is based on the theory of graph
separators. If the graphs considered are planar, a fundamental reference
is:

[1] R.J. Lipton and R.E. Tarjan, "A separator theorem for planar graphs",
     SIAM J. on Appl. Math., vol. 36, n. 2, pp. 177-189; April 1979.

This paper shows that an n-vertex planar graph can be separated into two
subgraphs of comparable sizes by removing at most $O(\sqrt{n})$ vertices, the
separator. Since it has been noted that digital circuits have properties
somehow similar to those of planar graphs, the above paper has motivated
a substantial amount of layout research. The separator theory was applied
to the layout of graphs in:

[2] C.E. Leiserson, "Area-efficient graph layouts (for VLSI)", Proc.21st
     IEEE Symp. on Fondations of Computer Science, Syracuse, NY, October
     1980; pp. 270-281.

and

[3] L.G. Valiant, "Universality considerations in VLSI circuits", IEEE
     Trans. on Computers, vol. C-30, n. 2, pp. 135-140; February 1981.

These (independently discovered) equivalent methods are based on the
divide-and-conquer principle. The separator theory is used to subdivide
the graph into two portions, these portions are recursively processed,
and the final layout is obtained by routing, with insignificant perturbation, the wires perturbation, the wires pertaining to the separator.

The important notions of "crossing number" and "wire area" of a
given graph, which are relevant to the layout area of the graph, were
introduced in

[4] F.T. Leighton, "New lower bound techniques for VLSI", Proc. 22nd
     IEEE Symp. on Foundations of Computer Science, Nashville, Tenn.,
     October 1981; pp. 1-12.

In this paper, Leighton also exhibited an n-vertex nonplanar graph--
the mesh of trees or orthogonal trees -- having a $\Theta(\sqrt{n})$-separator but
requiring $\Omega(n \log^2 n)$ layout area. This result can then be used to exhibit
an n-vertex planar graph -- the tree of meshes -- also with a $\Theta(\sqrt{n})$-
parator requiring $\Omega(n \log n)$ layout area. This shows a gap between lower
and upper bounds for planar graphs, which is still open today. Further
methods, which are applicable to arbitrary circuit graphs and are based
on the notion of "bifurcator", can be found in

[5] F.T. Leighton, "A layout strategy which is provably good", Proc.14th
     ACM Symp. on Theory of Computing, San Francisco, CA, May 1982; pp.
     85-98.

An interesting variation on the theory of separators (multicolor
separators), presented in

[6] J.R. Gilbert, "Graph separator theorems and sparse Gaussian Elimina-
     tion", Rep. N. STAN-CS-80-833, dept. of Comp. Sci., Stanford Uni-
     versity; December 1980.

has been successfully extended, and used in

[7] S.N. Bhatt and F.T. Leighton, "A framework for solving VLSI graph
     layout problems", Journal of Computer and System Sciences, to appear.

to prove the feasibility of "synchronous layouts". A synchronous layout
of a directed computation graph is realized when a node is laid out in an
area proportional to the total length of the wires corresponding to its
outgoing arcs. Bhatt and Leighton showed that an arbitrary n-node graph
laid out in area A, can be reprocessed to obtain a synchronous layout of
area at most $O(A \log^2 (A/N))$.

A large amount of research has been done on the *routing problem*,
defined by a fixed set of terminals on the uniform grid, the specification of their interconnection, and a region of the grid to be used to
realize the layout. A very early paper, which has been the basis of many
applications (especially for gate-array networks) is

[8] C.Y. Lee, "An algorithm for path connections and its applications",
     IRE Trans. on Computers, vol. EC-10, pp. 346-365; September 1961.

This paper assumes that module placement has been determined and successively routes the wires, each on the shortest feasible path on the grid.
The extreme simplicity of the technique - based on the propagation of a
"distance wave" - has been the reason for its success.

Another techniques presented in

[9] D.W. Hightower, "A solution to line routing problems on the conti-
     nuous plane", *Proc. 6th Design Automation Workshop*, pp. 1-24, June
     1969.

also aims at realizing wires as shortest paths (in the presence of bar-
riers) in the L_1-metric, but employs a different approach. Starting from
the L_1-shortest path, it verifies whether it crosses any barrier, and,if
so, it introduces approximate detours. A final compaction step removes
obvious redundancies.

Worth mentioning are the so-called "iterative" techniques, where
wires are laid out one after the other, until a wire is first found that
cannot be succesfully laid out: at this point, a small set of wires is
rerouted to allow the layout of the wire causing the impasse, and so on.
Typical of this approach are

[10] G.V. Dunn, "The design of printed circuit layouts by computer", *Proc.
     3rd Australian Computer Conf*. pp. 419-423, (1967).

[11] S.E. Lass, "Automated printed circuit routing with a stepping apert-
     ure", *Comm. of the ACM*, 12 n. 5, pp. 262-265, (1969).

The notion of "channel" was introduced in the paper

[12] A. Hashimoto and J. Stevens, "Wire routing by optimizing channel
     assignment within large apertures", *Proc. 8th Design Automation Work-
     shop*, pp. 155-169, June 1971.

which has become one of the classic references in layout theory. In the
channel routing problem, all terminals are on two parallel boards, whose
spacing (width) is determined by the number of tracks used by the layout.

A set of terminals to be connected together is called a *net*. Hashimoto and Stevens intriduced the notion of "density", which constitutes a lower bound to channel width under the usual assumption that the two distinct wires are edge-disjoint paths in the grid (no overlap).

Several papers have since addressed the channel routing problem (CRP). The following are two significant references:

[13] B.W. Kernighan, D.G. Schweikert, G. Persky, "An optimum channel routing algorithm for polycell layouts of integrated circuits", Proc. 10th Design Automation Workshop, pp. 50-59, June 1973.

[14] A. Deutsch, "A dogleg channel router", Proc. 13th Design Autmation Conference, pp. 425-433, 1976.

The first paper breaks away from usual approach of applying a promising heuristic, and obtains an optimal solution (i.e., with the least number of tracks) by a branch and bound technique: of course running time is greatly affected by the problem size. The second paper presents an interesting method to reduce the number of tracks used. All the CRP techniques cited so far refer to the so-called Manhattan-mode (or two-layer) routing, where horizontal and vertical wires are on distinct conducting layers, with appropriate cuts (vias) established to provide the necessary contacts.

A significant step forward is represented by the paper

[15] R.L. Rivest, A. Baratz, and G. Miller, "Provably good channel routing algorithms", Proc. 1981 Carnegie-Mellon Conf. on VLSI, pp. 153-159, October 1981.

The adopted layout mode is the knock-knee (earlier used by Thompson), where two distinct wires are allowed to share a bend-point. They prove that a two-terminal net CRP of density d can be laid out in d tracks in the knock-knee mode; however, since only two condicting layers are postulated, additional d-1 tracks are introduced to provide the necessary vias. The width 2d-1 was later found to be optimal for two-layer routing, as shown in:

[16] F.T. Leighton, "New lower bounds for channel routing", draft 1981.

Since the use of two layers appears to be the determining factor of
the 2d-1 width performance (rather than d), the paper

[17] F.P. Preparata and W. Lipski,Jr., "Three layers are enough", Proc.
23rd IEEE Symp. on Foundations of Computer Science, Chicago, IL,
pp. 350-357, November 1982 (see also: Preparata-Lipski, "Optimal
three-layer channel routines", IEEE Trans. on Computers, May 1984
(to appear)),

showed indeed that for a two-terminal net CRP it is possible to produce
a minimal width layout, which is wireable in no more than three layers.
This paper is also the basis of the wireability theory to be briefly
mentioned below.

The routing of multiterminal nets, earlier approached on the basis
of reasonable heuristics and the subject of intensive experimentation
[13,14], has only recently received theoretical attention. Although
density trivially represents a lower bound to channel width for this
general CRP, the establishment of the optimum width in an NP-hard problem,
at least in the Manhattan mode, as shown in:

[18] T.G. Szymanski, "Dogleg channel routing is NP-complete", to appear
(1982).

However, in an unpublished memorandum

[19] D.J. Brown, F.P. Preparata, "Three-layer routing of multiterminal
nets", unpublished manuscript, October 1982,

an upper bound of 2d to the achievable channel width was established.
This upper bound was later improved to (2d-1) in the paper

[20] M. Sarrafzadeh and F.P. Preparata, "Compact channel routing of multiterminal nets", Tech. Rep. ACT. 44, Coordinated Science Lab., University of Illinois, October 1983.

The algorithm reported in this paper produces the layout column by column
in a left-to-right sweep of the channel, and falls in the general class
of "greedy" channel routers. An experimentally efficient, but not yet

analyzed, greedy router for the Manhattan mode was described in the paper.

[21] R.L. Rivest, C.M. Fiduccia, "A greedy channel router", Proc. 19
     Design Automation Conference, pp. 418-424, June 1982.

This paper is part of the 'PL' system, presently being developed at M.I.T.
An account of this layout system, encompassing placement and routing, can
be found in

[22] R.L. Rivest, "The 'PI' (Placement and Interconnect) System", Proc.
     19th Design Automation Conference, pp. 475-481, June 1982.

The problem of the number of layers used to realize a given layout
is currently the subject of active research. The general theoretical
framework presented in [17] has been used in

[22] W. Lipski, Jr., "The structure of three-layer wireable layouts", to
     appear in Advances in Computing Research, Volume 2, VLSI Theory,
     (1984),

to prove that the problem of deciding whether an arbitrary planar layout
of multiterminal nets is wireable in three layers is NP-complete. However, this result has been supplemented by the surprising and elegant
finding that for the same problem no more than four layers are ever necessary, as reported in

[23] M. Brady and D.J. Brown, "VLSI routing: four layers suffice", in
     Advances in Computing Research Volume 2: VLSI Theory, (1984).

Finally we mention some recent results on routing problems of a
more general flavor than CRPs. The first problem, discussed in

[24] A.S. LaPaugh, "A polynomial time algorithm for optimal routing
around a rectangle", Proc. 21st Symp. on Foundations of Computer Science
(Syracuse), pp. 282-293, October 1980.

[25] T.F. Gonzales and S.L. Lee, "An optimal algorithm for optimal routing
around a rectangle", Proc. 20th Allerton Conference on Communication
Control, and Computing, pp. 636-645, October 1982,

concerns the construction of the layout when the terminals are placed on the four sides of a rectangle and the wires must remain external to this rectangle ("Routing outside a rectangle"). The second - strictly related - problem, for the same data, prescribes instead that the layout be constructed inside the rectangle ("Routing inside a rectangle"). The general theory of the latter problem for two-terminal nets was first presented in

[26] A. Frank, "Disjoint paths in a rectilinear grid", Combinatorica, 2, 4, pp. 361-371, (1982),

and an efficient algorithm was later developed in

[27] K. Mehlhorn and F.P. Preparata, "Routing through a rectangle", Tech. Rep. ACT-42, Coordinated Science Lab., Univ. of Illinois, Urbana: October 1983; submitted for publication.

These generalized routing problems for multiterminal nets have so far received scant attention.

In conclusion, it emerges from this selected bibliography that, in spite of extensive studies and of significant heuristic accomplishmed the combinatorial understanding of the planar layout problem is still a preliminary stage. Although most problems are, or are likely to be intractable, there is a strong need for the development of fully analyzed approximation methods.

Printed in the United States
by Bookmasters

Printed in the United States
By Bookmasters